Molded Thermosets

SPE Books

Ulrich; *Introduction to Industrial Polymers*

Saechtling; *International Plastics Handbook for the Technologist, Engineer and User*

Stoeckhert; *Mold-Making Handbook for the Plastics Engineer*

Bernhardt; *Computer Aided Engineering for Injection Molding*

Michaeli; *Extrusion Dies—Design and Engineering Computations*

Rauwendaal; *Polymer Extrusion*

Brostow/Corneliussen; *Failure of Plastics*

Menges/Mohren; *How to Make Injection Molds*

Throne; *Thermoforming*

Manzione; *Applications of Computer Aided Engineering in Injection Molding*

Macosko; *Fundamentals of Reaction Injection Molding*

Tucker; *Fundamentals of Computer Modeling for Polymer Processing*

Charrier; *Polymeric Materials and Processing—Plastics, Elastomers and Composites*

Wright; *Molded Thermosets: A Handbook for Plastics Engineers, Molders and Designers*

RALPH E. WRIGHT

MOLDED THERMOSETS

A Handbook for Plastics Engineers, Molders, and Designers

HANSER

Hanser Publishers,
Munich Vienna New York Barcelona

Distributed in the United States of America and in Canada by
Oxford University Press, New York

Ralph E. Wright
Ralph E. Wright Associates
48 Pleasant Street
Yarmouth, Maine 04096

Distributed in the USA and in Canada by
Oxford University Press
200 Madison Avenue
New York, NY 10016

Distributed in all other countries by
Carl Hanser Verlag
Kolbergerstrasse 22
D-8000 München 80

Library of Congress Cataloging-in-Publication Data

Wright, Ralph E.
Molded thermosets : a handbook for plastics engineers, molders, and designers / Ralph E. Wright.
p. cm.
Includes index.
ISBN 0-19-520920-6 (Oxford University Press)
1. Thermosetting plastics. I. Title.
TP1180.T55W75 1991
668.4′22—dc20 91-73380
CIP

Die Deutsche Bibliothek – CIP-Einheitsaufnahme

Wright, Ralph E.:
Molded thermosets : a handbook for plastics engineers, molders, and designers / Ralph E. Wright. – Munich ; Vienna ; New York ; Barcelona : Hanser, 1991
ISBN 3-446-15820-0

ISBN 3-446-15820-0 Carl Hanser Verlag, Munich, Vienna, New York, Barcelona
ISBN 0-19-520920-6 Oxford University Press

Text Typesetting: Publishers Service of Montana Inc., Bozeman, Montana
Printed in Germany

To my wife, Jane, without whose steadfast support and encouragement, this book could never have been started or finished.

Foreword

The Society of Plastics Engineers (SPE) is pleased to sponsor and endorse this new volume, *Molded Thermosets*. The author, Ralph E. Wright, has been an instructor in SPE's Educational Seminar Program for a number of years. His course, "Thermosets: Applications, Materials and Processes," has served as the base on which this book was developed.

Thus, the volume makes available at a substantially reduced cost and in a far more extensive format information previously imparted only to seminar registrants. This broad base provides a background for an understanding of such specialized topics as cost estimating, product and mold design, and molding techniques. Certainly, a significant contribution is the last chapter, which looks into the crystal ball of the future.

SPE, through its Technical Volumes Committee, has long sponsored books on various aspects of plastics. Its involvement has ranged from identification of needed volumes to recruitment of authors and peer review, and approval of new publications.

Technical competence pervades all SPE activities, not only in the publication of books, but also in other areas such as sponsorship of technical conferences and educational programs. In addition, the Society publishes periodicals including, *Plastics Engineering*, *Polymer Engineering and Science*, *Polymer Processing and Rheology*, *Journal of Vinyl Technology* and *Polymer Composites*, as well as conference proceedings and other publications, all of which are subject to rigorous technical review procedures.

The resource of some 36,000 practicing plastics engineers has made SPE the largest organization of its type worldwide. Further information is available from the Society at 14 Fairfield Drive, Brookfield, Connecticut 06804, U.S.A.

Robert D. Forger
Executive Director
Society of Plastics Engineers

Technical Volumes Committee

Robert E. Nunn, Chairman
University of Lowell

Preface

In the early part of the twentieth century, Dr. Leo Baekeland (1907) achieved a reaction between phenol, a caustic, crystalline acidic compound, and formaldehyde, a colorless, pungent gas, by heating these two ingredients under pressure in a reaction kettle. The resultant, hot liquid, when cooled, became a hard, brittle, and frangible solid. It lacked physical strength, but possessed excellent thermal, chemical, and electrical insulating properties. This hard, brittle resin in a pulverized form could then be mixed with certain fillers, or reinforcements, colorants, lubricants, and catalysts. The end product was a molding compound capable of further processing on heated roll mills and calendering mills. It could also be passed through granulators to achieve a specified particle size and distribution. These operations were carried out under carefully controlled time and temperature conditions.

Thus was the first thermosetting molding compound created. When this compound was fed into a heated mold and thereby subjected to sufficient pressure to cause the material to move or flow to form itself into the configuration of the mold cavities, it underwent an irreversible chemical reaction now known as "polymerization," or curing. The molded part was permanently transformed. In addition, all of its mechanical, electrical, chemical, and thermal properties were enhanced, resulting in the creation of a new material whose property profile included excellent heat resistance, superior electrical insulation, easy moldability, low cost, and most importantly, adaptability to mass-production techniques.

This discovery by Baekeland brought about a new look at synthetic resins. As a result, such resins were first recognized by their ability to achieve this "polymerized" state, with its attendant beneficial properties. Resin systems capable of such cure were designated as "thermosets," whereas all other resin systems became known as "thermoplastics."

In the ensuing years following Baekeland's work, additional thermoset resin systems were developed to meet the requirements of diverse applications in the automotive, communication, construction, transportation, household appliance, business machine, aircraft, and aerospace fields.

It is the purpose of this text book to present, in an orderly format, all phases of the thermosetting molding industry from the raw materials to the finished molded article. The goal is to enhance the understanding of this field to those already involved in the industry, to students contemplating a career in thermosets, as well as to design and manufacturing engineers desirous of furthering their understanding of benefits to be derived from the use of these versatile materials. Although many of the thermosetting molding compounds may be prepared for liquid casting and potting applications, such preparations and techniques will not be presented in this volume.

Ralph E. Wright

Acknowledgments

There can be little doubt about the value of having knowledgeable friends when the time comes to embark on the task of writing a book. The author would be remiss in not acknowledging the assistance and courtesies of a goodly number of men and women who have been of immense help during the writing of this text.

Several Rogers Corporation personnel both at MMD and at Corporation headquarters, including Walter Hayes, Lorraine Murphy, David Daniels, Mark DeLassus, Ray Mikulak, Craig McLea, Michael Gardner, Dr. Vincent Landi, Bruce Fitts, and Cindy Duksta, lent a helping hand whenever it was requested.

The faculty at the University of Lowell's Plastic Engineering School, headed by Dr. Nick Schott and including Professors Stephen Orroth, Robert Nunn, and Stephen Driscoll, all provided technical as well as personal support in great measure.

David M. Wright has been extremely helpful, not only with his insight and experience as a Plastics Engineer, but also with his deep knowledge of computer technology, without which this book would have never seen the publisher's desk.

Among the many firms that provided technical assistance as well as product information for inclusion in this book were: American Cyanamid Co., Hysol Div. Dexter Corp., ICI/Fiberite, Finmac Incorporated, Plastics Engineering Co., Hull Corp., Pennwalt Corp. Bucher, Dusan Equipment Corp. Neu Dynamics, Werner & Pfleiderer Corp. Buss (America) D-M-E Corp. Pre-Mix Inc. Dow Corning Corp.

Guest Foreword

It has been said that the best way to expand the quality of life is the proper and synergistic application of all types of materials.

And, if that statement seems rather idealistic, and lacks a compelling motivation to better understand the properties and applications of engineering materials, consider this motivation: Survival in a fiercely competitive global market!

Yes, growing competition in a globalized market has characterized the past decade. It has required new thinking and new directions in product design and application of materials. It has also required a restructuring of manufacturing methods and attitudes.

Survivors will be world class companies which are innovative, dynamic, and positioned for rapid and continuous improvement. These world class organizations will be focused on serving customers' needs with creative, high quality products, just in time, and at the lowest cost.

This textbook will focus on a single, but increasingly important, class of materials—thermoset plastic molding compounds. These materials are ideally positioned to help designers and companies not only to survive, but to actually achieve those world class objectives. Thermoset molding compounds generally provide outstanding dimensional stability under heat and load, combined with exceptionally high-flexural modulus and surface hardness. They are readily moldable into intricate configurations in very rapid cycles, making them highly cost effective design materials.

Thermoset plastics' thermal stability is due to their rigid, highly cross-linked molecular structure when molded and cured. Designers require that their parts retain their design dimensions and performance characteristics throughout their expected life service and operating environment, but at the lowest cost. In many instances, designing with thermosetting compounds can prove to be the ideal solution.

Thermoset plastic resin systems contain a high percentage and variety of complementary reinforcement materials, permitting many combinations of mechanical, thermal, electrical, chemical, and other important engineering properties. Designers can take advantage of these properties to optimize performance in any particular application.

Thermosets, of course, have greatly benefited from the high-tech tools available for predicting performance, optimizing part configurations, improving properties, and consistency. With our increased understanding of how compounds behave, thermosets are continuing to be viewed as primary design materials.

However, thermoset molded components are usually hidden from the view of the ordinary observer—inside a computer, deep down inside a mechanism where they are most likely to be in a high performance operating environment—being subjected to prolonged periods of elevated temperatures and pressure, standing off corrosive forces and maybe even withstanding centrifugal forces of thousands of RPMs or possibly providing electrical insulation within minute tolerances where dimensional instability would mean failure.

Some typical illustrations would include electric motor commutators, aerospace connectors, torque converter reactors, brake pistons, pulleys, water pumps, coil bobbins, intake manifolds, and so on.

The traditional thermoset resin systems most commonly used today in molding compounds are allyls, alkyds, phenolics, epoxies, thermoset polyesters, ureas, melamines, polyurethanes, and silicones. These resin systems cover a considerable balance of properties, individually, or as a group, including thermal and dimensional stability, electrical insulation, chemical resistance, high-flexural modulus, toughness, surface hardness, moldability, and cost effectiveness.

Each of these resin systems has its own distinct advantages depending on applicational requirements. However, each system also has tremendous potential in terms of resin alloying, coupled with a broad selection of reinforcement or filler systems. The molding compounds can be tailored and fine tuned to meet the design requirements of a broad range of specific applications.

Read on for the keys to understanding and using the revitalized engineering materials of the future. Notwithstanding the impressive advances in thermoplastic materials over the past years, thermoset plastic molding compounds may well hold a key to your product's performance and competitiveness in the world marketplace of tomorrow.

Walter A Hayes
Vice President
Rogers Corporation

Contents

CHAPTER 1

THERMOSETTING MOLDING COMPOUNDS

The family of thermosetting molding compounds currently in wide use consists of seven members, each one offering its own unique set of properties for consideration by the product designer, the raw material supplier, or the molder.

The seven members are

Allyls (DAP, DAIP)
Aminos (Urea/Melamine)
Epoxies
Phenolics
Thermoset Polyesters
Silicones
Composites*

The intent of this chapter is to present an orderly review of each family member with detailed information on their characteristics such as

Chemistry
Specifications
Reinforcements
Data Sheet Values
Applications

*Increased use of thermosetting resins in conjunction with a wide variety of new reinforcing materials, molding techniques, and application areas has led us to composites—a relatively new member of the thermosetting industry that will be covered under Section 1.7.

1.1 ALLYLS (DAP, DAIP)

The allyl resins used in the manufacture of diallyl phthalate molding compounds are either the orthophthalate (DAP) or the isophthalate (DAIP) prepolymers, with the ortho (DAP) accounting for approximately 90% of the total usage. The fundamental difference between the two resin systems lies in the fact that the iso (DAIP) resin will cure faster at molding temperature and has about 50°C higher heat resistance, but its cost is significantly higher. The compounds utilizing the ortho (DAP) resin are generally regarded as superior to any other of the thermosets currently available in their market niche.

Compounds based on these resin systems are available in a wide variety of reinforcement grades and are readily moldable in all conventional thermosetting molding techniques. Articles molded with these compounds will exhibit superior electrical, mechanical, chemical, and thermal properties, which they retain even when subjected to severe environmental excursions of elevated temperature and humidity when under load and over long time spans.

Chemistry

The diallyl phthalate monomer is an ester produced by the esterification process involving a reaction between an acid (phthalic anhydride) and an alcohol (allyl alcohol) as shown below:

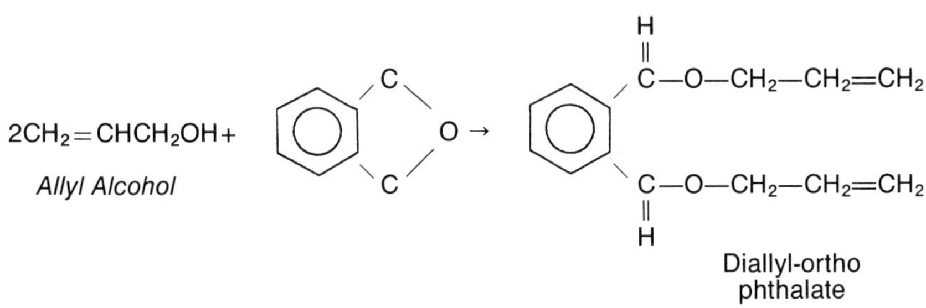

The monomer is capable of being cross-linked and polymerized in the presence of certain peroxide catalysts such as

- Dicumyl peroxide (DICUP)
- *t*-Butyl perbenzoate (TBP)
- *t*-Butylperoxyisopropyl carbonate (TBIC)
- 2,5-Dimethyl-2,5-bis(benzoylperoxy)hexane (Luperox 118)
- Ethyl-3,3-bis(*t*-butylperoxy butyrate) (USP 333)
- α,α′-Bis(*t*-butylperoxy)diisopropylbenzene (Vul-cup)
- 1,1-Bis(*t*-butylperoxy)cyclohexane (USP 400 P)

Catalyzed compounds possess excellent shelf life when stored in sealed containers at normal room temperatures. DAIP compounds with very soft flow characteristics should be stored at somewhat lower temperatures (40–50°F) (4.4–10°C). The curing mechanism in either system proceeds very rapidly during the molding process, where the compounds encounter higher temperatures (300–375°F, 149–191°C), resulting in cure times as low as 20 sec, depending on the molded part thickness and general configuration and the ability to employ molding temperatures at the highest possible degree.

Specifications

DAP and DAIP molding compounds meet the requirements of several commercial and military specifications as well as those of certain industrial and electronic firms who will use military specifications in conjunction with their own special requirements.

Military

Mil-M-14 is the primary thermosetting compound specification, and it lists both DAP and DAIP compounds under the following designations:

MDG—Mineral filler. General purpose.
SDG—Short glass filler. General purpose.
SDG-F—Short glass filler. Flame resistant.
SDI-5—Orlon filler. Impact value 0.5 ft-lb.
SDI-30—Dacron filler. Impact value 3.0 ft-lb.
GDI-30—Long glass filler. Impact value 3.0 ft-lb.
GDI-30F—Long glass filler. Impact value 3.0 ft-lb. Flame resistant.
GDI-300—Long glass filler. Impact value 30 ft-lb.
GDI-300F—Long glass filler. Impact value 30 ft-lb. Flame resistant.
Note: All of the above grades contain the ortho (DAP) resin system.

The latest revision of Mil-M-14 (H) lists designations for those compounds using the isophthalate (DAIP) resin system and refers to them as *heat resistant* compounds:

MIG—Mineral filler. Heat resistant.
MIG-F—Mineral filler. Heat resistant.
SIG—Short glass filler. Heat resistant.
SIG-F—Short glass filler. Heat resistant. Flame resistant.
GII-30—Long glass filler. Heat resistant. Impact 3.0 ft-lb.
GII-30F—Long glass filler. Heat/flame resistant. Impact value 3.0 ft-lb.
Note: All of the above DAIP compounds will perform without substantial property loss during long-term exposure to elevated temperatures of 490°F (200°C).

Table 1. Typical data sheet values: Allyls

		Grade and Mil-M-14 type					
Property[a]	ASTM test method	51-01CAFR MDG	775 CAF MDG	FS-6 CAF MDG	52-01 SDG	FS-5 SDG	52-70-70 VO SDG-F
Reinforcement	—	Mineral	Mineral	Mineral	Short glass	Short glass	Short glass
Resin-isomer (DAP)	—	Ortho	Ortho	ISO	Ortho	ISO	Ortho
Form	—	Granular	Granular	Granular	Granular	Granular	Granular
Bulk factor	—	2.3	2.8	2.3	2.3	2.3	2.3
Specific gravity (g/cm^3)	D-792A	1.75	1.67	1.86	1.93	1.91	1.91
Shrinkage (comp) in/in.	D-955	0.0050	0.0065	0.0035	0.002–0.004	0.002–0.004	0.002–0.004
Izod impact (ft/lb/in.)	D-256A	0.4	0.4	0.4	0.6	0.6	0.6
Flexural strength (psi)	D-790	10000	10000	10500	12000	11000	14000
Flexural modulus (psi) $\times 10^6$	D-790	1.2	1.0	1.3	1.9	2.0	2.0
Tensile strength (psi)	D-638	—	—	—	6500	6500	6500
Compressive strength (psi)	D-695	—	—	—	—	—	—
Water absorption %, 24 hr at 100°C/48 hr at 50°C in H_2O	D-570	0.55	0.65	0.50	0.25	0.35	0.25
CTE (in./in./°C) $\times 10^{-5}$	D696	—	—	—	—	—	—
Deflection temperature (°F)	D-648A	300–350	300–350	300–350	500	500	500
Flammability (ign./burn) sec	Fed. 2023	90 90	—	—	—	—	90A 90
U.L. flammability rating							
1/8 in.	U.L. 94	94V-O	—	—	—	—	94V-O
1/16 in.	U.L. 94	94V-O	—	—	—	—	94V-O
Comparative tracking index (sec)	—	600 +	600 +	600 +	600 +	600 +	600 +
Dielectric strength 60 Hz ST/SS wet (VPM)	D-149	375	400	350	350	360	350
Dielectric constant 1 kHz/1 MHz wet	D-150	4.0	4.5	4.1	4.5	4.6	4.4
Dissipation factor 1 kHz/1 MHz wet	D-150	0.037	0.023	0.035	0.016	0.014	0.014
Arc resistance (sec)	D-495	135	123	175	150	175	130

Table 1. *(Continued)*

		Grade and Mil-M-14 type						
Property[a]	ASTM test method	73-70-70 C SDG-F	73-70-70 R SDG-F	FS-10 VO SDG-F	52-20-30 GDI-30	FS-4 GDI-30	52-40-40 GDI-30F	FS-80 GDI-30F
Reinforcement	–	Short glass	Short glass	Short glass	Long glass	Long glass	Long glass	Long glass
Resin-isomer (DAP)	–	Ortho	Ortho	ISO	Ortho	ISO	Ortho	ISO
Form	–	Granular	Granular	Granular	Flake	Flake	Flake	Flake
Bulk factor	–	2.4	2.3	2.3	6	6	6	6
Specific gravity (g/cm^3)	D-792A	1.91	1.85	1.91	1.72	1.64	1.79	1.74
Shrinkage (comp) in/in.	D-955	0.002–0.004	0.002–0.004	0.002–0.004	0.0025	0.0025	0.0025	0.0025
Izod impact (ft/lb/in.)	D-256A	0.6	0.6	0.6	4.0	3.6	3.9	3.4
Flexural strength (psi)	D-790	13000	14000	13000	15000	15000	15000	15000
Flexural modulus (psi) $\times 10^6$	D-790	2.0	1.9	2.0	1.3	1.4	1.3	1.4
Tensile strength (psi)	D-638	7000	6500	8000	–	–	–	–
Compressive strength (psi)	D-695	–	–	–	–	–	–	–
Water absorption %, 24 hr at 100°C/48 hr at 50°C in H_2O	D-570	0.25	0.30	0.35	0.35	0.35	0.32	0.35
CTE (in./in./°C) $\times 10^{-5}$	D-696	–	–	–	–	–	–	–
Deflection temperature (°F)	D-648A	500	450	500	500	500	500	500
Flammability (ign./burn) sec	Fed. 2023	90A 90	90A 90	90A 90	–	–	90A 90	90A 90
U.L. flammability rating								
1/8 in.	U.L. 94	94V-0	94V-0	94V-0	–	–	–	–
1/16 in.	U.L. 94	94V-0	94V-0	94V-0	–	–	–	–
Comparative tracking index (sec)	–	600 +	600 +	600 +	600 +	600 +	600 +	600 +
Dielectric strength 60 Hz ST/SS wet (VPM)	D-149	350	350	350	350	350	350	340
Dielectric constant 1 kHz/1 MHz wet	D-150	4.4	4.2	4.6	4.2	4.0	4.3	4.1
Dissipation factor 1 kHz/1 MHz wet	D-150	0.014	0.015	0.016	0.016	0.014	0.016	0.015
Arc resistance (sec)	D-495	130	125	175	130	138	133	145

Table 1. *(Continued)*

		Grade and Mil-M-14 type						
Property[a]	ASTM test method	RX 1-503 SDI-5	RX 1-501N MDG	RX 1-501AN MDG	RX 1-510N MDG	RX 2-501N MDG	RX 3-1-501N MDG	RX 1-540 —
Reinforcement	—	Orlon	Mineral	Mineral Nylon	Mineral	Mineral	Mineral	Short-glass mineral
Resin-isomer (DAP)	—	Ortho	Ortho	Ortho	Ortho	ISO	Ortho	Ortho
Form	—	Granular	Granular	Granular	Granular	Granular	Granular	Granular
Bulk factor	—	3.0	2.3	2.6	2.4	2.1	2.0	2.3
Specific gravity (g/cm^3)	D-792A	1.40	1.80	1.73	1.66	1.80	1.83	1.85
Shrinkage (comp) in/in.	D-955	0.010–0.012	0.003–0.006	0.007–0.010	0.003–0.006	0.003–0.006	0.003–0.006	0.001–0.003
Izod impact (ft/lb/in.)	D-256A	0.65	0.45	0.45	0.45	0.45	0.60	0.60
Flexural strength (psi)	D-790	11000	16000	15000	14000	16000	16000	18000
Flexural modulus (psi) $\times 10^6$	D-790	0.7	2.2	1.3	1.6	2.2	2.2	2.2
Tensile strength (psi)	D-638	5000	8000	7500	7500	8000	8000	8000
Compressive strength (psi)	D-695	29000	24000	25000	23000	25000	25000	23000
Water absorption %, 24 hr at 100°C/48 hr at 50°C in H_2O	D-570	0.40	0.30	0.40	0.50	0.25	0.30	0.35
CTE (in./in./°C) $\times 10^{-5}$	D696	2.2	2.5	3.4	2.9	2.3	2.1	1.9
Deflection temperature (°F)	D-648A	290	330	300	325	525	325	425
Flammability (ign./burn) sec	Fed. 2023	—	—	—	—	—	105 25	—
U.L. flammability rating								
1/8 in.	U.L. 94	—	94HB	—	—	94HB	94V-0	94HB
1/16 in.	U.L. 94	—	94HB	—	—	94HB	94V-0	94HB
Comparative tracking index (sec)	—	—	—	—	—	—	—	—
Dielectric strength 60 Hz ST/SS wet (VPM)	D-149	375/350	400/375	410/375	400/375	400/375	400/375	400/375
Dielectric constant 1 kHz/1 MHz wet	D-150	0.4/3.1	4.2/3.6	4.1/3.7	4.2/3.3	4.3/3.3	4.3/3.7	4.4/4.2
Dissipation factor 1 kHz/1 MHz wet	D-150	0.016/0.021	0.013/0.014	0.000/0.010	0.015/0.017	0.015/0.019	0.013/0.016	0.013/0.018
Arc resistance (sec)	D-495	125	140	135	135	150	135	180

Table 1. *(Continued)*

Property[a]	ASTM test method	RX 2-520 SDG	RX 1-520 RX 1310 SDG	RX 3-2-520F SDG-F	RX 3-1-525F RX 1366FR SDG-F	RX-1-530 GDI-30	RX-2-530 GDI-30	RX 3-1-530 GDI-30F	RX 2-530 GDI-30F
		Grade and Mil-M-14 type							
Reinforcement	–	Short-glass	Short-glass	Short-glass	Short-glass	Long-glass	Long-glass	Long-glass	Long-glass
Resin-isomer (DAP)	–	Ortho	Ortho	Iso	Ortho	Ortho	Iso	Ortho	
Form	–	Granular	Granular	Granular	Granular	Flake	Flake	Flake	Flake
Bulk factor	–	2.5	2.4	2.4	2.3	3.5	3.5	3.5	
Specific gravity (g/cm^3)	D-792A	1.75	1.83	1.90	1.87	1.74	1.73	1.76	1.00
Shrinkage (comp) in/in.	D-955	0.001–0.003	0.001–0.003	0.001–0.003	0.001–0.003	0.001–0.004	0.001–0.004	0.001–0.004	0.000–0.000
Izod impact (ft/lb/in.)	D-256A	0.80	0.70	0.80	0.80	3.0–7.0	3.0–7.0	3.0–7.0	3.0–0.0
Flexural strength (psi)	D-790	22000	22000	23000	23000	23000	23000	23000	23000
Flexural modulus (psi) × 10^6	D-790	2.6	2.6	2.6	2.6	1.1	2.0	2.0	0.0
Tensile strength (psi)	D-638	12000	11000	12000	12000	10000	10000	10000	10000
Compressive strength (psi)	D-695	25000	25000	25000	25000	27000	27000	27000	27000
Water absorption %, 24 hr at 100°C/48 hr at 50°C in H_2O	D-570	0.25	0.25	0.25	0.25	0.35	0.35	0.35	0.00
CTE (in./in./°C) × 10^{-5}	D-696	1.7	1.5	1.2	1.5	1.6	1.5	2.2	0.0
Deflection temperature (°F)	D-648A	525	400	525	400	500	550	500	550
Flammability (ign./burn) sec	Fed. 2023	–	–	1120 40	110 40	–	–	110 40	110 00
U.L. flammability rating									
1/8 in.	U.L. 94	94HB	–	94V-O	94V-O	–	–	94V-O	–
1/16 in.	U.L. 94	94HB	–	94V-O	94V-O	–	–	94V-O	–
Comparative tracking index (sec)	–	–	–	600 +	600 +	–	–	600 +	–
Dielectric strength 60 Hz ST/SS wet (VPM)	D-149	450/400	450/400	450/400	450/400	400/375	400/375	400/375	400/375
Dielectric constant 1 kHz/1 MHz wet	D-150	3.9/3.7	4.0/3.5	4.1/3.9	4.2/3.5	0.0/0.0	3.8/3.6	4.1/3.5	4.1/0.0
Dissipation factor 1 kHz/1 MHz wet	D-150	0.011–0.015	0.009–0.016	0.010–0.013	0.010–0.016	0.011–0.019	0.010–0.019	0.010–0.018	0.010–0.019
Arc resistance (sec)	D-495	150	130	150	130	135	135	130	135

[a]Certification to Mil-M-14 requires batch testing. All testing in accordance with Mil-M-14. Values are typical and not statistical minimums.

ASTM

Under D-1636-75A, the DAP and DAIP molding compounds are listed by type, class, and grade as follows:

Type	Class	Grade
I	A. Ortho (DAP) B. Ortho FR[a] (DAO) C. Iso (DAIP) D. Iso (DAIP)	1. Long glass 2. Medium glass 3. Short glass
II	A. Ortho (DAP) B. Ortho FR (DAP) C. Iso (DAIP) D. Iso FR (DAIP)	1. Mineral 2. Mineral/organic fiber
III	A. Ortho (DAP)	1. Acrylic fiber 2. Polyester fiber long 3. Polyester fiber milled

[a]FR, flame resistant.

U.L. Listing

Underwriters Laboratory (U.L.) currently rates DAP and DAIP thermally at 266°F (130°C) with flame resistance rating of 94 VO in 1/16th in. section. Neither DAP nor DAIP compounds have been submitted to U.L.'s long-term thermal testing. However, Mil-M-14 (H) does recognize their excellent long-term heat resistance and ability to retain their initial property profile under adverse thermal conditions over lengthy time spans.

Reinforcements

As is noted in both the MIL and ASTM specifications, the DAP and DAIP compounds are available with a broad range of fillers or reinforcements:

- Minerals
- Organic fibers
- Cotton flock
- Acrylic fibers
- Short and long glass fibers
- Polyethylene-teraphthalate fibers

Typical Data Sheet Values[1]

Typical values are given in Table 1. They all display the properties summarized below:

Summary of Properties of Allyls
Physical
Excellent long-term dimensional stability
Virtually no postmold shrinkage
Chemically inert
Mechanical
Excellent flexural strength
High impact resistance
Highly resistant to sudden, extreme jolts and severe stresses
Electrical
Retains high insulation resistance at elevated temperatures
Retains performance characteristics at high ambient humidity
Thermal
Successful performance in vapor-phase soldering environments—419° and 487°F (215° and 253°C)
FS-10VO:200°C U.L. temperature index
Isophthalate materials are recommended for very high temperature applications
Chemical
Highly resistant to solvents, acids, alkalies, fuels, hydraulic fluids, plating chemicals, and sterilizing solutions

Applications

Due to their outstanding property profile and the ability to retain those properties even when subjected to severe environmental conditions involving elevated temperatures, high ambient humidity under load over time has led to their use by designers in those application areas in which unfailing performance is essential, such as

- Commercial and military electrical connectors
- Terminal boards
- Circuit breaker housings
- Potentiometer housings

- X-Ray tube housings
- Computer panel boards
- Rotary switches
- Potting cups
- Underwater cable components
- Relays
- Dental equipment (high temperature sterilization)

Suppliers U.S.A.

All of the current DAP/DAIP molding compound suppliers in the United States have complete lines of products that can be molded in all of the current thermoset molding processes and are available in a wide range of colors, flows, and granulations. The U.S. suppliers are

- Rogers Corp., MMD Div., Manchester, CT 06040
- Cosmic Plastics Inc., San Fernando, CA 91342-5381
- Occidental Chemical, Durez Div., N. Tonawanda, NY 14120-3493

1.2 AMINOS

The amino resins are the basis of two primary molding compound types:

- Urea—Formaldehyde
- Melamine—Formaldehyde

Ureas

The ureas are generally compounded with wood flour as their chief reinforcement, providing a free-flowing granular compound that can easily be preformed, preheated, or preplasticated. The compounds can also be readily molded with both compression and transfer molding techniques but need special formulations when processed with screw injection equipment due to the need to extend their flow "life" at the higher barrel temperatures.

Melamines

Melamine molding compounds, unlike the ureas, have a wide range of grades utilizing wood flour, minerals, and/or α-cellulose fibers as their reinforcements. These will result in free-flowing compounds suitable for all premolding preparations as well as molding in all of the conventional thermosetting molding processes. In addition, melamine compounds are also available in the higher strength, higher bulk factor compounds using glass fibers and/or chopped cotton flock as reinforcement. These compounds can be molded in compression and transfer molding but will require the use of auxiliary devices to overcome their lack of pourability if employed with the screw injection process.

Chemistry

Both urea and melamine will react with formaldehyde, initially by addition, forming methylol compounds, and then by condensation in reactions similiar to those occurring with phenol and formaldehyde.[2]

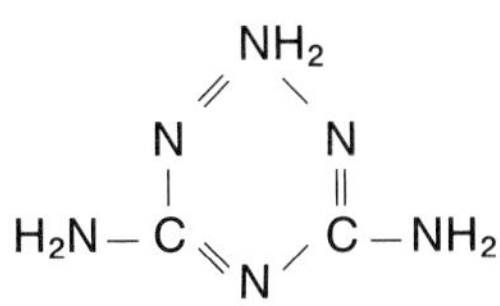

Melamine. The methyol reaction appears as

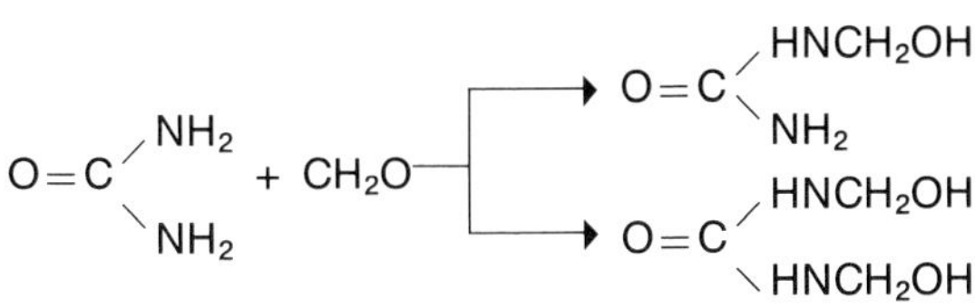

These methyol compounds can polymerize to cross-linked resins by the loss of water, while other reactions that may enter into the condensation include the formation of methyleneurea:

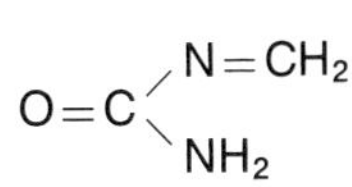

with the following cross-linking reactions:

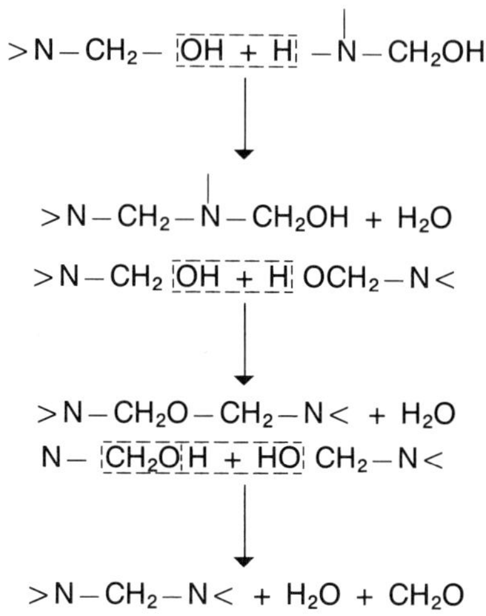

Specifications

Military
Mil-M-14:

Urea: not listed.
Melamine: Types:

CMG—Cellulose filler. General purpose.
CMI-5—Cellulose filler. Impact value 0.5 ft-lb.
CMI-10—Cellulose filler. Impact value 1.0 ft-lb.
MMD—Mineral filler. General purpose.
MMI-5—Glass filler. Impact value 0.5 ft-lb.
MMI-30—Glass filler. Impact value 3.0 ft-lb.

U.L. Listing

	Thermal index	Flame resistance
Urea	221°F (105°C) max.	94 VO in 1/16th-in. section
Melamine	220–292°F (130–150°C)	94 VO in 1/16th-in. section

Reinforcements

- Purified α-cellulose fibers
- Minerals
- Chopped cotton flock

- Wood flour
- Glass fibers (long and short)

Typical Data Sheet Values[3]

Melamines: See Table 2.
Ureas: See Table 3.
Melamine/Phenolics: See Table 4.[4]

Table 2. Typical data sheet values: Melamine

	Filler				
	α Cellulose	Mineral	Chopped cotton fabric	Cellulose	Glass
Specific gravity	1.49	1.78	1.5	1.5	1.97
Flex strength (psi)	10,000	7,600	12,000	7,000	10,500
Impact strength (ft-lb)	0.25	0.3	0.6	0.25	3.0–4.0
Water absorption % gain (24 hr at 25°C)	0.4	0.15	0.5	0.65	0.2
Dielectric strength (V/M)					
Short time	300	350	250	350	170
Step × step	250	250	100	200	170
Arc resistance (sec)	120	120	125	70	180

Table 3. Typical data sheet values: Ureas

	Filler	
	α Cellulose	Cellulose
Specific gravity	1.49	1.49
Flex strength (psi)	8000	7500
Impact strength (ft-lb)	0.2–0.3	0.2–0.3
Water absorption % gain (24 hr at 25°C)	0.6	0.6
Dielectric strength (V/M)		
Short time	350	350
Step × step	275	275
Arc resistance (sec)	120	120

Table 4. Typical data sheet values: Melamine/Phenolics

	Properties[a]	ASTM test method	00713 Colors	00714 Brown	00732 Colors	00755 Brown	000757 Brown
*C	Compound						
*T	Specific gravity	D-792	1.55	1.67	1.64	1.60	1.65
*I			1.55	1.67	1.64	1.60	1.65
			–	–	–	1.60	1.70
	Apparent density	D-1895	0.66	0.65	0.60	0.56	0.60
	Bulk factor	D-1895	2.3	2.5	2.8	2.9	2.8
	Preforming		Good	Good	Good	Good	Good
	Storage life (months)		3	3	3	3	3
C	Mold shrinkage	D-955	0.006	0.006	0.006	0.006	0.006
T	(½ × ½ × 5 in. bar)		0.009	0.009	0.009	0.009	0.009
I			–	–	–	0.008	0.010
C	Deflection temperature	D-648	285	330	305	330	335
T	(°F) at 264 psi		325	350	350	355	345
I			–	–	–	270	340
C	Water absorption (%)	D-570	0.4	0.4	0.45	0.42	0.35
T			0.4	0.3	0.30	0.3	0.25
I			–	–	–	0.2	0.25
C	Heat resistance (°F)	D-794	300	325	300	325	325
T			300	325	300	325	325
I			–	–	–	325	325
	Mechanical (compression)						
	Izod impact strength (units)	D-256	0.29	0.26	0.34	0.32	0.36
	Tensile strength (psi)	D-638	6,200	6,400	6,500	7,900	6,500
	Flexural strength (psi)	D-790	8,500	9,800	8,500	9,900	9,300
	Modulus in flexure (psi)	D-790	1.1×10^6	1.3×10^6	1.3×10^5	1.2×10^6	1.3×10^6
	Compressive strength (psi)	D-695	28,500	24,000	26,500	25,500	24,000
	Mechanical (transfer)						
	Izod impact strength (ft-lb/in.)	D-256	0.27	0.26	0.33	0.30	0.34
	Tensile strength (psi)	D-638	9,000	9,000	9,000	9,000	8,400
	Flexural strength (psi)	D-790	10,800	12,000	10,900	12,000	12,600
	Modulus in flexure (psi)	D-790	1.1×10^6	1.3×10^6	1.8×10^6	1.1×10^6	1.2×10^6
	Compressive strength (psi)	D-695	29,000	24,500	27,000	25,700	25,500
	Mechanical (injection)						
	Izod impact strength (ft-lb/in.)	D-256				0.40	0.34
	Tensile strength (psi)	D-638				7,800	8,000
	Flexural strength (psi)	D-790				12,000	12,000
	Modulus in flexure (psi)	D-790				1.0×10^6	1.2×10^6
	Compressive strength (psi)	D-695				22,000	23,000
C	Electrical		275	300	275	300	300
T	Dielectric strength v/mil (S.T.)	D-149	250	240	260	275	250

Table 4. *(Continued)*

	Properties[a]	ASTM test method	00713 Colors	00714 Brown	00732 Colors	00755 Brown	000757 Brown
C	Dissipation factor (%)		3.3	5.4	4.2	5.1	5.5
T	1 MHz	D-150	4.0	5.3	4.8	7.4	7.0
C	Permitivity		6.0	7.2	6.7	7.0	7.0
T	1 MHz	D-150	7.0	7.2	7.2	7.6	7.5
C	Volume resistivity	D-257	4.5×10^{12}	4.5×10^{12}	4.5×10^{12}	4.5×0^{12}	4.5×10^{12}
T			4.5×10^{12}	4.5×10^{12}	4.5×10^{12}	4.5×10^{12}	4.5×10^{12}
	Arc resistance	D-495	130	180	130	182	182
	Comparative tracking index	D-3638	400	400	400	500	600
	Additional properties						
	Drop ball impact	PLENCO	9	9	16	13	13
	Dielectric breakdown at 250°F (kV)	PLENCO					
	Post shrinkage 72 hr at 250°F	PLENCO					
	Surface resistivity	D-257					
	Designed to meet Mil-M-14 type ASTM D-700 type		CMG	CMG	CMG	CMG	CMG

[a]Resin type.
*C = Compression molded; T = Transfer molded; I = Injection molded.

Applications

Ureas Articles molded with the urea compounds will display a high gloss, scratch resistant surface that is very desirable in applications such as

- Closures
- Control housings
- Wiring devices
- Control buttons
- Electric shaver housings
- Knobs

These applications take full advantage of the wide range of light-fast colors available with the urea compounds.

Melamines Molding compounds utilizing the melamine resin system have applications in the commercial, industrial, and military areas where their reinforcement variety, scratch resistance, and wide color range are very useful. Typical commercial applications include

- Molded dinnerware
- Electric shaver housings
- Knobs
- Buttons
- Ashtrays
- Connector bodies

Typical military applications:

- Connector bodies
- Circuit breaker housings

Combinations of phenolic and melamine resins in molding compounds have produced compounds with excellent color stability and ease of moldability and are used in heat resistant applications such as

- Appliance handles
- Pot and pan handles

Suppliers U.S.A.

Amino compounds:

- American Cyanamid Co., Polymer Prod. Div., Wayne, NJ 07470
- ICI/Fiberite, 501 W. 3rd St., Winona, MN 55987

Melamine/phenolic compounds:

- Plastics Engineering Co., P.O. Box 758, Sheboygan, WI 53038

1.3 EPOXIES

The epoxy resin-based molding compounds serve two distinct market areas:

- Military electronic and electrical applications requiring high strength, dimensional stability, and excellent electrical properties at elevated temperatures
- Commercial and military electronic applications utilizing encapsulation molding techniques

Chemistry

The liquid resin requires the addition of a hardener to achieve total cross-linking. The high performance encapsulation compounds consist of an epoxy-novolac resin with amine catalysts whereas the lower performance compounds utilize the bisphenol A–epichlorohydrin system with anhydride catalysts.

Specifications

Military

Mil-M-24325 (ships) 1968: Types:

MEC—Mineral filler. Encapsulation grade. Low pressure.
MEE—Mineral filler. Electrical grade. Low pressure.
MEG—Mineral filler. General purpose.
MEH—Mineral filler. Heat resistant.
GEI-5—Glass filler. Impact value 0.5 ft-lb.
GEI-20—Glass filler. Impact value 2.0 ft-lb.
GEI-100—Glass filler. Impact value 10.0 ft-lb.

U.L. Listing

Epoxy molding compounds carry a thermal index rating of 266°F (130°C) with a flammability rating of 94 VO.

ASTM

Listed under D-3013-77.

Reinforcements

- Minerals
- Glass fibers (long and short)
- Fused or crystalline silica

Typical Data Sheet Values[5]

Typical values are given in Table 5.

Applications

Due to their retention of electrical, thermal, mechanical, and chemical properties at elevated temperatures and their very high moisture resistance, the high-

Table 5. Typical data sheet values: Epoxies

Property[a]	ASTM test method	Class						
		Hardware grade						
		1904B	1906	2004B	1907	1908 1908B	1914	2008
Reinforcement		Mineral glass	Mineral glass	Mineral glass	Short-glass	Short-glass	Short-glass	Short-glass
Bulk factor	—	2.4	2.3	2.3	2.2	2.3	2.1	2.5
Form	—	Granular	Granular	Granular	Granular	Granular	Granular	Granular
Specific gravity (g/cm^3)	D-792	2.05	1.90	1.95	1.95	1.85	1.94	1.85
Mold shrinkage (in./in.)	D-955	0.001–0.003	0.001–0.003	0.002–0.004	0.002–0.004	0.002–0.004	0.001–0.003	0.003–0.005
Impact strength (ft/lb/in.)	D-256	0.61	0.40	0.60	0.58	0.55	0.61	0.62
Flexural strength (RT,psi.)	D-790	17000	15000	16000	15000	17000	17000	16000
Flexural modulus (psi $\times$ 10^6)	D-790	1.9	2.1	1.8	1.9	1.9	2.2	1.9
Tensile strength (psi)	D-651	10000	10000	10000	10000	9000	10500	9000
Compressive strength (psi)	D-695	32000	32000	32000	32000	33000	31000	32000
Water absorption (%), 48 hr at 122°F (50°C)	D-570	0.2	0.2	0.2	0.3	0.3	0.3	0.2
Barcol hardness	D-2583	72	70	72	73	72	70	72
Deflection temperature (°F)	D-648	500	340	500	450	500	500	500
Coefficient of linear thermal expansion (in./in./°C $\times$ 10^{-6}) (−30°C–+30°C)	D-696	28	37	33	47	37	41	37
Coefficient of thermal conductivity (cal/sec/cm^3/°C/cm $\times$ 10^{-4})	—	17.5	24	14.5	14	17.5	14	16.5

Dielectric constant								
(1 MHz/24 hr at 23°C H_2O)	D-150	4.5	4.8	4.7	4.8	4.8	5.0	4.6
Dissipation factor								
(1 MHz/24 hr at 23°C H_2O)	D-150	0.02	0.02	0.02	0.02	0.03	0.02	0.02
Dielectric strength (S/S)								
(VPM/48 hr at 50°C H_2O)	D-149	300	330	300	325	300	300	325
Arc resistance (sec)	D-495	186	185	182	180	184	185	180
Insulation resistance								
(MΩ/30 days at 150°F, 95% RH)	D-257	10^1	10^1	10^1	10^1	10^1	10^1	10^1
Oxygen index	D-2863	37	43	50	38	40	39	50
U.L. flammability rating								
1/4 in.	U.L. 94	—	—	—	—	94V-O	94V-O	—
1/16 in.	U.L. 94	—	—	94V-O	—	—	—	—
Mil-M-14, type	—	—	MFH	—	MFH	—	—	—

Table 5. *(Continued)*

Property[a]	ASTM test method	Class			
		Electrical encapsulation grade			
		1960B	1961B	2060B	2061B
Reinforcement		Mineral glass	Mineral glass	Mineral glass	Mineral glass
Bulk factor	—	2.5	2.3	2.5	2.3
Form	—	Granular	Granular	Granular	Granular
Specific gravity (g/cm^3)	D-792	1.75	1.90	1.80	1.95
Mold shrinkage (in./in.)	D-955	0.004–0.006	0.002–0.001	0.004–0.006	0.002–0.001
Impact strength (ft/lb/in.)	D-256	0.56	0.63	0.60	0.61
Flexural strength (RT,psi.)	D-790	15000	14000	15000	14000
Flexural modulus (psi $\times 10^6$)	D-790	1.7	2.0	1.7	2.0
Tensile strength (psi)	D-651	9000	10500	9000	11000
Compressive strength (psi)	D-695	30000	31500	30500	31000
Water absorption (%), 48 hr at 122°F (50°C)	D-570	0.3	0.2	0.3	0.3
Barcol hardness	D-2583	65	65	65	70
Deflection temperature (°F)	D-648	350	425	375	450
Coefficient of linear thermal expansion (in./in./°C $\times 10^{-6}$) (−30°C–+30°C)	D-696	38	38	35	42
Coefficient of thermal conductivity (cal/sec/cm^3/°C/cm $\times 10^{-4}$)	—	14	16	17	13

Dielectric constant					
(1 MHz/24 hr at 23°C H_2O)	D-150	4.2	4.5	4.0	4.2
Dissipation factor					
(1 MHz/24 hr at 23°C H_2O)	D-150	0.02	0.02	0.02	0.02
Dielectric strength (S/S)					
(VPM/48 hr at 50°C H_2O)	D-149	325	325	320	320
Arc resistance (sec.)	D-495	185	186	170	170
Insulation resistance					
(MΩ/30 days at 150°F, 95% RH)	D-257	10^1	10^1	10^1	10^1
Oxygen index	D-2863	50	3208	—	45.0
U.L. flammability rating					
1/4 in.	U.L. 94	—	—	—	—
1/16 in.	U.L. 94	—	—	—	94V-O
Mil-M-14, type	—	—	—	—	

[a]Certification to Mil-M-14 requires batch testing. All testing in accordance with Mil-M-14. Values are typical and not statistical minimums.

pressure epoxy molding compounds listed under Mil-M-24325 have found a market in military and commercial high-performance applications such as

- Electrical connectors
- Potting shells
- Relays
- Printed circuit board cores
- Switches
- Coils
- Bobbins

The low-pressure molding compounds that exhibit the same high-performance characteristics as the high-pressure compounds have become the primary insulating medium for encapsulating components such as

- Integrated circuits
- Resistors
- Coils
- Diodes
- Capacitors
- Relays
- Bobbins

The ever expanding use, worldwide, of computers and computer-driven electronic devices and systems with their heavy use of encapsulated components provides an extremely active market base for the low-pressure epoxy encapsulation molding compounds. This market is currently in the 75–85 mm lb range annually.

Suppliers U.S.A.

Rogers Corp. MMD, Manchester, CT 06040
Dexter Electronic Materials Div., 211 Franklin St., Olean, NY 14760
ICI/Fiberite, 501 W. 3rd St., Winona, MN 55987
Morton Chemical, 333 W. Wacker Dr., Chicago IL 60606-1292
Plaskon Electronic Materials Inc., 2829 Glendale Ave., Toledo, OH 43614

1.4 PHENOLICS

The phenolic resin-based molding compounds are, by far, the most widely known and used of all the thermosets. As was noted in the Preface to this book, Dr. Leo Baekeland's successful reaction between phenol and formaldehyde in the early years of this century became the foundation for the entire family of thermosetting molding compounds.

Over past 72 years, phenolics have found their niche in commercial, industrial, and military applications due to their inherent property profile, diverse range of reinforcements, low cost, and extreme ease of molding in all of the conventional thermosetting molding processes. Even today, these versatile compounds are the primary choice of product designers for applications involving elevated temperatures under load.

Chemistry

Phenols react with aldehydes to give condensation products if there are free positions on the benzene ring—*ortho* and *para* to the hydroxyl group. Formaldehyde is by far the most reactive and is used almost exclusively in commercial production. The reaction is always catalyzed, either by acids or by bases. The nature of the product is greatly dependent on the type of catalyst and the mole ratio of the reactants.

There are four major reactions in the phenolic resin chemistry, with the novolac (two stage) and the resole (single stage) being the two used for the production of molding compounds.

Novolac Resins

In the presence of acid catalysts, and with the mole ratio of formaldehyde to phenol less than 1, the methylol derivatives condense with phenol to form, first, dihydroxydiphenyl methane[6]:

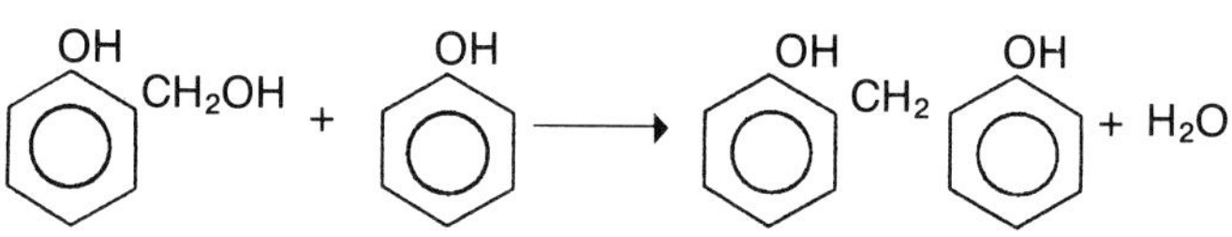

and, on further condensation and methyl bridge formation, fusible and soluble linear low polymers called "novlacs" with the following structure[6]:

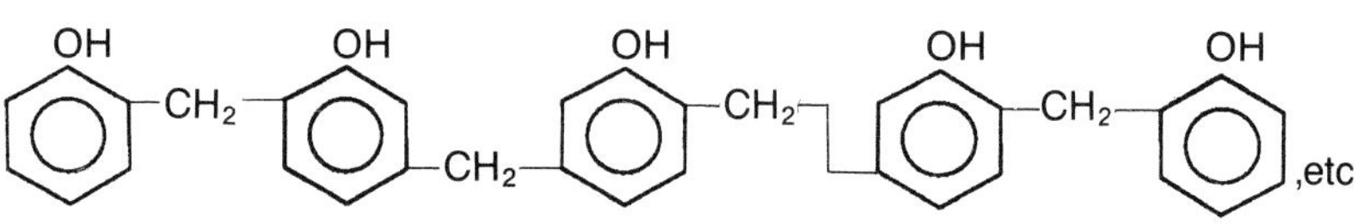

where the *ortho* and *para* links occur at random.

The novolac (two stage) resins are made with an acid catalyst, and only part of the necessary formaldehyde is added to the reaction kettle, producing a mole ratio of about 0.8:1. The rest is added later as hexamethylenetetramine, which decomposes in the final curing step, with heat and moisture present, to yield formaldehyde and ammonia, which act as the catalyst for curing.

Resole Formation

In the presence of alkaline catalysts and with more formaldehyde, the methylol phenols can condense either through methylene linkages or through ether linkages. In the latter case, subsequent loss of formaldehyde may occur with methylene bridge formation[7]:

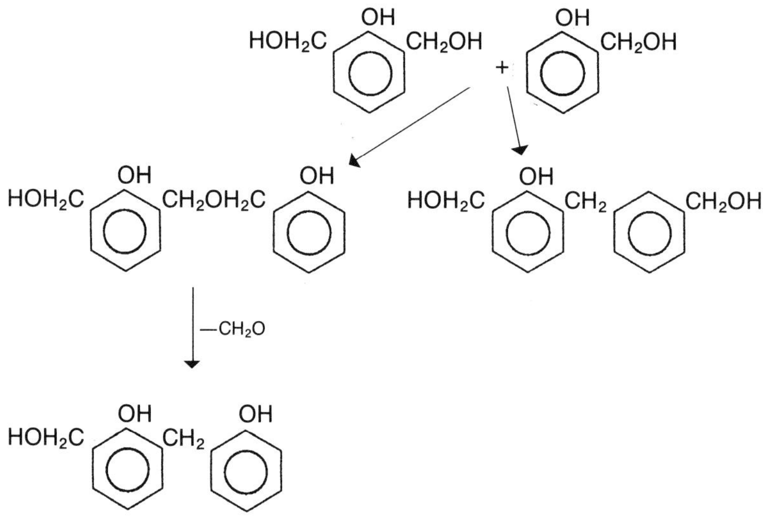

Products of this type, soluble and fusible but containing alcohol groups, are called "resoles." If the reactions leading to their formation are carried further, large numbers of phenol nuclei can condense to give a network formation.

In the production of a single stage phenolic resin, all the necessary reactants for the final polymer (phenol, formaldehyde, and catalyst) are charged into a resin kettle and reacted together.

The ratio of formaldehyde to phenol is about 1.25:1, and an alkaline catalyst is used.

The resole (single stage) resin based molding compounds will not outgas ammonia during the molding process as do the novolac (two stage) compounds. This leads the resole compounds to be used for applications where their lack of outgassing is most beneficial. Their chief drawback is that they are not easy to control during their production because they are susceptible to the effects of temperatures in excess of 45°F/7.2°C. Production of these compounds is usually scheduled for the cooler months of the year, and they must be stored in sealed containers in areas where the temperature is maintained at the 45°F/7.2°C maximum limit.

The novolac resin-based compounds are the most widely used despite the ammonia outgassing, which can be taken care of with proper mold venting. They have an outstanding property profile and are easily molded in all known thermosetting molding methods. Their chief drawback to more extensive use is their limited color range—basically black, brown, and dark green.

Reinforcements

- Wood flour (refined sawdust)
- Cotton flock
- Minerals—clay/mica/calcium carbonate
- Chopped fabric
- Short and long glass fibers
- Teflon
- Nylon
- Rubber
- Kevlar

Typical Data Sheet Values[7]

Typical values are given in Table 6. They all display the properties summarized below.

Summary of Properties of Phenolics
Physical
Excellent dimensional stability under thermal cycling and high stress conditions
Low water absorption
High surface hardness
Mechanical
High compressive strength
High modulus (stiffness)
RX660C is designed for higher strength in thin walled sections
Thermal
Phenolic retains a high percentage of its properties at elevated temperatures
Low coefficient of thermal expansion
Glass-transition temperatures of over 450°F (232°C) are attainable
Ammonia-free phenolics are excellent candidates for vapor-phase soldering
Chemical
Highly resistant to automotive fluids (glycol/petrochemicals/hydrocarbons), solvents, and some acids and bases

Table 6. Typical data sheet values: Phenolics

Property[a]	ASTM test method	Cellulose reinforced				
		CR-1				
		RX 340	RX 342	RX 475	RX 431	RX 525
Phenolic resin type	—	2-Step	2-Step	2-Step	2-Step	2-Step
Form	—	Pellet	Pellet	Pellet	Pellet	Pellet
Bulk factor	—	2.2	2.3	3.3	3.5	2.6
Specific gravity (g/cm^3)	D-792A	1.43	1.44	1.37	1.42	1.43
Shrinkage (Comp.)(in./in.)	D-955	0.006	0.006	0.006	0.005	0.006
Izod impact (ft/lb/in.)	D-256A	0.50	0.55	2.0	1.1	0.55
Flexural strength (psi)[b]	D-790	10000	10000	12000	12000	10000
Flexural modulus ($psi \times 10^6$)	D-790	1.2	1.2	1.3	1.6	1.3
Tensile strength (psi)	D-638	6000	6000	5500	6000	6500
Compressive strength (psi)	D-695	25000	23000	23000	27000	26500
Water absorption (%),						
24 hr at 50°C	D-570	1.0	1.0	2.1	2.0	1.5
24 hr at 23°C	D-570	0.1	0.1	0.35	0.25	0.1
Deflection Temperature (°F)						
64 psi	D-648A	350	325	330	325	325
Temperature index (°C)[c]	—	—	—	—	150	150
Flammability rating						
1/8 in.	U.L. 94	—	—	—	94HB	94V0
1/16 in.	U.L. 94	—	—	—	94HB	94HB
Coefficient of thermal						
expansion (in./in./°C $\times 10^{-5}$)	D-696	5.0	5.0	3.0	5.0	5.2
Dielectric strength						
Hz, ST/SS, wet (VPM)	D-149	—	—	60/50	60/50	150/125
Arc resistance (sec)	D-495	—	—	—	100	125
Mil-M1-14 Type	—	—	—	CFI-20	CFI-10	CFG

Table 6. *(Continued)*

Property[a]	ASTM test method	Glass reinforced GR-1 RX 630	RX 611	RX 660	RX 660C	RX 670	RX 643	RX 640	RX 647
Phenolic resin type	–	2-Step	2-Step	2-Step	2-Step	2-Step	1-Step	1-Step	1-Step
Form	–	Granular	Granular	Granular	Granular	Granular	Granular	Granular	Granular
Bulk factor	–	2.1	2.1	2.1	2.1	2.0	2.1	2.0	2.0
Specific gravity (g/cm^3)	D-792A	1.78	1.75	1.80	1.82	1.90	1.74	1.74	1.90
Shrinkage (Comp.)(in./in.)	D-955	0.0015	0.002	0.002	0.002	0.002	0.0015	0.002	0.002
Izod impact (ft/lb/in.)	D-256A	0.90	0.80	0.70	0.70	0.65	1.0	0.90	0.70
Flexural strength (psi)	D-790	31500	27500	20000	22000	20000	30000	27000	22000
Flexural modulus ($psi \times 10^6$)	D-790	3.2	3.0	2.9	3.0	3.0	2.7	2.5	2.9
Tensile strength (psi)	D-638	12000	12500	10000	9000	8000	13000	12500	10500
Compressive strength (psi)	D-695	38500	37500	32000	33000	37000	35000	32000	35000
Water absorption (%),									
24 hr at 50°C	D-570	0.20	0.20	0.20	0.25	0.20	0.20	0.20	0.20
24 hr at 23°C	D-570	0.02	0.05	0.02	0.02	0.02	0.08	0.08	0.04
Deflection Temperature (°F)									
64 psi	D-648A	380	375	375	400	400	530	530	530
Temperature index (°C)	150	150	180	180	150	150	–	150	–
Flammability rating									
1/8 in.	U.L. 94	94V0	94V0	94V0	94V0	94V0	–	94V0	–
1/16 in.	U.L. 94	94V0	94V0	94V0	94V0	94V0	–	94V0	–
Coefficient of thermal expansion (in./in./°C $\times 10^{-5}$)	D-696	1.4	1.5	1.8	1.8	1.6	1.3	1.3	1.4
Dielectric strength									
Hz, ST/SS, wet (VMP)	D-149	400/350	400/350	350/300	350/300	350/300	350/300	350/300	350/300
Arc resistance (sec)	D-495	180	180	180	180	180	180	180	180
Mil-M1-14 Type	–	–	MFH	–	–	–	GPI-10	GPI-5	GPI-5

Table 6. *(Continued)*

	ASTM test	Glass reinforced					
		GR-2					
Property[a]	method	RX 865	RX 868	RX 867	RX 866	RX 853	HT 500[d]
Phenolic resin type	—	2-Step	2-Step	2-Step	2-Step	2-Step	—
Form	—	Flake	Pellet	Pellet	Pellet	Pellet	Flake
Bulk factor	—	3.0	2.5	2.6	3.0	3.0	3.1
Specific gravity (g/cm^3)	D-792A	1.86	1.77	1.77	1.70	1.80	1.75
Shrinkage (Comp.) (in./in.)	D-955	0.002	0.002	0.0025	0.002	0.002	0.002
Izod impact (ft/lb/in.)	D-256A	1.5	0.70	0.65	0.65	0.65	1.0
Flexural strength (psi)	D-790	19000	14000	14000	12500	14000	13500
Flexural modulus ($psi \times 10^6$)	D-790	3.2	2.3	2.5	2.3	2.7	2.0
Tensile strength (psi)	D-638	8000	6000	7000	8000	8000	8500
Compressive strength (psi)	D-695	35000	19000	21000	27000	27000	25000
Water absorption (%),							
24 hr at 50°C	D-570	0.20	0.35	0.35	0.20	0.35	0.10
24 hr at 23°C	D-570	0.04	0.16	0.17	0.05	0.18	0.01
Deflection Temperature (°F) 64 psi	D-648A	525	480	375	400	380	550 +
Temperature index (°C)		180	—	150	—	150	[d]

Flammability rating								
1/8 in.	U.L. 94	94V0	—	94V0	—	94V0	94V0
1/16 in.	U.L. 94	94V0	—	94V0	—	94V0	94V2
Coefficient of thermal expansion							
(in./in./°C × 10^{-5})	D-696	1.3	—	—	—	—	—
Dielectric strength							
Hz, ST/SS, wet (VMP)	D-149	350/250	175/125	175/125	200/150	375/250	250/150
Arc resistance (sec)	D-495	180	180	180	180	180	180
Mil-M1-14 Type	—	—	—	—	—	—	—

[a]Certification Mil-M-14 requires batch testing. All testing in accordance with Mil-M-14. Values are typical and not statistical minimums.
[b]Connection factor = 1 psi = 0.0068947 MPa.
[c]U.L. temperature index evaluations are presently underway.
[d]HT-500 is a special thermoset alloy. It is characterized by an elevated temperature life that is up to 5 times longer than typical phenolic materials with U.L. temperature indexes of 180°C.

Specifications

Phenolic molding compounds are recognized under three major molding compound specifications.

Military

Mil-M-14: Types:

CFG—Cellulose filler. General purpose.
CFI-5—Cellulose filler. Impact value 0.5 ft-lb.
CFI-10—Cellulose filler. Impact value 1.0 ft-lb.
CFI-20—Cellulose filler. Impact value 2.0 ft-lb.
CFI-30—Cellulose filler. Impact value 3.0 ft-lb.
CFI-40—Cellulose filler. Impact value 4.0 ft-lb.
MFE—Mineral filler. Best electrical grade.
MFH—Mineral filler. Heat resistant.
GPG—Glass filler. General purpose.
GPI—Glass filler. Impact value 0.5 ft-lb.
GPI-10—Glass filler. Impact value 1.0 ft-lb.
GPI-20—Glass filler. Impact value 2.0 ft-lb.
GPI-30—Glass filler. Impact value 3.0 ft-lb.
GPI-40—Glass filler. Impact value 4.0 ft-lb.

ASTM

Listed under ASTM D700-65.

U.L. Listings

Phenolic molding compounds were among the very first thermosets to have received U.L. recognition and mostly retain a thermal index rating of 155°C, but there are a number of newer glass filled compounds as well as some phenolic alloys that have received thermal ratings in the 185°C range. Most phenolic compounds carry a flame retardant rating of 94 VO in a 1/16th-in. section.

Applications

The ability to combine the phenolic resins with a broad range of reinforcement materials has contributed to their versatility, allowing design engineers real flexibility in their material selection process.

Figure 1. Transmission stator/reactor.

This, coupled with their low cost, excellent mechanical, electrical, chemical, and thermal properties, moldability, and superb dimensional stability at elevated temperature when under load, has led to their worldwide acceptance in such diverse market areas as

- Automotive
- General transportation
- Electronics
- Aeronautics
- Aerospace
- Electrical
- Appliances
- Business equipment

(see Figures 1–4).[8]

Suppliers U.S.A.

Rogers Corp. MMD, Manchester, CT 06040
ICI/Fiberite, 501 W. 3rd St., Winona, MN 55987

Figure 2. Water pump housing.

Plastics Engineering Co., Sheboygan, WI 53082
Plaslok Corp., Buffalo, NY 14227
Occidental Chemical, Durez Div., N. Tonawanda, NY 14120-3493
Resinoid Engineering Corp., Mat. Div., Skokie, IL 60076
Valentine Sugars Inc., Valite Div., Lockport, LA 70374

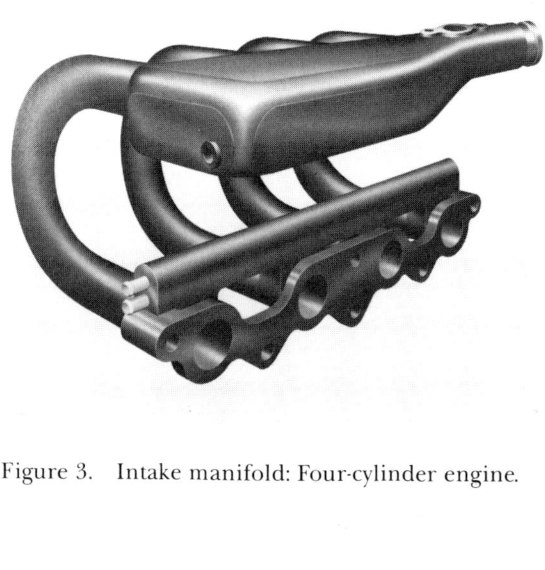

Figure 3. Intake manifold: Four-cylinder engine.

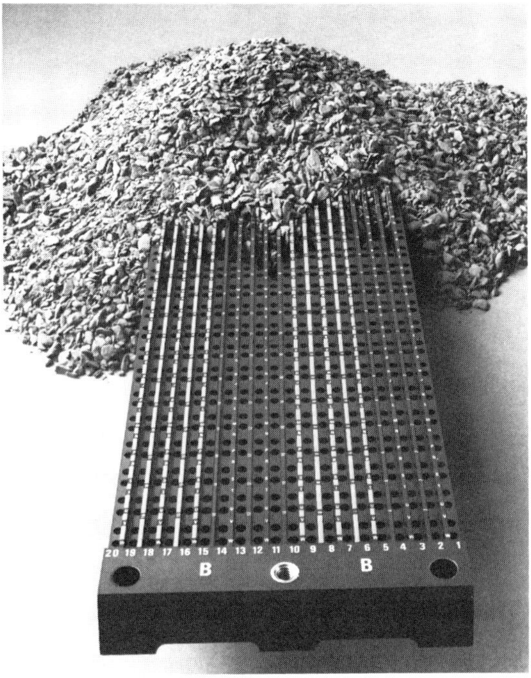

Figure 4. Glass-filled phenolic molding compound terminal board.

1.5 THERMOSET POLYESTERS

The so-called "thermoset polyesters" have been making rapid strides over the past decade due to their low cost, wide color selection, moldability, and good electrical properties, along with the fact that they are available in a variety of physical forms—free-flowing granular, pelletized (PMG), putty or rope-type extrudates, sheet molding compounds (SMC), high bulk factor compounds (BMC), and thick molding compounds (TMC).

Chemistry

Thermosetting polyesters can be produced from phthalic or maleic anhydrides and polyfunctional alcohols. Catalyzation is generally achieved by the use of free radical producing peroxides—not too unlike the DAP/DAIP compounds. Actually, unsaturated DAP and/or vinyl toluene monomers are frequently added to formulations to promote flow and to extend their flow life. Cross-linking occurs through the free-radical initiated polymerization of maleic and fumaric double bonds, or by copolymerization with added unsaturated monomers. Since no

Table 7. Typical data sheet values: Thermoset polyesters

Sheet molding compounds (typical ASTM data)	ASTM test method	U.L. recognized electrical		
		4000-( )-( )	42000-( )-( )	47000-( )-( )
Water absorption (%)	D-570	.2	.18	.15
Specific gravity	—	1.85	1.75	1.85
U.L. 94 flame classification — U.L. File E-27875	—	94V-0	94V-0	94V-0, 945-V
Tensile strength (psi)	D-638	14,000	10,000	15,000
Flexural strength (psi)	D-790	30,000	25,000	30,000
Compressive strength, flatwise (psi)	D-695	30,000	28,000	30,000
Impact strength, Izod, edgewise (ft/lb/in. notch)	D-256	13.5	18.5	20.0
Arc resistance (sec)	D-495	180	180	180
Track resistance (min)	D-2303	450	>600	500
Dielectric strength, perpendicular, short time, VPM in oil	D-149	420	425	420
Dielectric constant at 60 Hz	D-150	4.7	4.4	4.4
Dissipation factor at 60 Hz	D-150	.96	.96	.95
Glass content, percent for data shown	—	22	22	33
Range of glass content available (%)	—	15-35	15-35	33
Mold shrinkage (in./in.)	—	.002	.001	.002

Table 7. *(Continued)*

Bulk molding compounds (typical ASTM data)	ASTM test method	U.L. recognized electrical 13600-( )-( )	14100-( )-( )	17000-( )-( )
Water absorption (%)	D–570	.35	.1	.10
Specific gravity	–	1.94	1.89	1.9
U.L. 94 flame classification – U.L. File E-27875	–	94V-0	94V-0	94V-0, 945-V
Tensile strength (psi)	D–638	6,000	6,000	7,000
Flexural strength (psi)	D–790	15,000	12,500	15,000
Compressive strength, flatwise (psi)	D–695	15,000	13,000	20,000
Impact strength, Izod, edgewise (ft/lb/in. notch)	D–256	3.5	3.5	5.0
Arc resistance (sec)	D–495	190	192	180
Track resistance (min)	D–2303	>600	>600	600
Dielectric strength, perpendicular, short time, VPM in oil	D–149	375	400	400
Dielectric constant at 60 Hz	D–150	5.8	5.0	5.0
Dissipation factor at 60 Hz	D–150	295	3.25	3.0
Glass content, percent for data shown	–	15	15	20
Range of glass content available (%)	–	5-35	5-35	20
Mold shrinkage (in./in.)	–	.002	.0001	.002

volatile by-products are evolved during cure, these materials can be fabricated by low-pressure molding with very short molding cycles.

Specifications

Military

Mil-M-14. The alkyd group is listed as follows:

MAG—Mineral filler. General purpose.
MAI-30—Glass filler. Impact value 3.0 ft-lb.
MAI-60—Glass filler. Impact value 6.0 ft-lb.
MAT-30—Glass filler. Impact value 3.0 ft-lb.

The MAT-30 is rated as flame and track resistant.

ASTM

Covered under D-1201-80 and D-1201-62 (reapproved 1975).

U.L. Listing

Thermoset polyester compounds generally carry a 356°F (180°C) thermal index rating with a flammability rating of 94 VO in a 1/16-in. section.

Reinforcements

- Minerals
- Long and short glass fibers
- Organic fibers

Typical Data Sheet Values

Typical values are given in Table 7.

Applications

Due to their versatility, color range, electrical properties, and low cost the thermoset polyesters have found their market niche in the following applications:

- Automotive components
- Large circuit breaker housings
- Brush holders
- Commercial connectors
- Battery racks
- Business machine housings

Suppliers U.S.A.

Rogers Corp. MMD, Manchester, CT 06040
American Cyanamid Co., Polymer Prod., Wayne, NJ 07470
Premix Inc., N. Kingsville, OH 44068
Applied Composites, St. Charles, IL 60174
ICI/Fiberite, 510 W. 3rd St., Winona, MN 55987
Haysite Reinforced Plastics, Erie, PA 16509
Plastics Engineering Co., Sheboygan, WI 53082-0758
Glastic Co., Cleveland, OH 44121

1.6 SILICONES

Silicone molding compounds utilize a base resin that is the result of reacting silicone with methylchloride to produce methylchlorosilanes.

A typical molding compound will contain 20–25% resin (phenyl and methyl siloxanes), 75% filler (mixtures of glass fibers and fused silica), lead-based catalysts, and pigments and lubricants that contribute flow promotion and good mold release.

The monomer is prepared by a process such as

$$Si + RCl \underset{\text{heat}}{\overset{\text{catalyst}}{\rightarrow}} RSiCl_3 + R_2SiCl_2, \text{ etc.}$$

Specifications

Military

Mil-M-14: Types:

MSG—Mineral filler. Heat resistant.
MSI-30—Glass filler. Impact value 3.0 ft-lb. Heat resistant.

U.L. Listing

Thermal index rating of 464°F (240°C)
Flammability rating 94 VO in 1/16-in. section

Table 8. Typical data sheet values: Silicones

	Dow Corning silicone molding compound					Test method	
Properties[a]	306	307	308	480	1-5021	CTM[b]	ASTM
Physical							
Specific gravity	1.88	1.85	1.86	1.92	1.92	0540	D–792
Flexural strength (psi × 10^3)	8.8 ± 0.7	7.5 ± 0.8	9.2 ± 1.4	9.1 ± 1.1	8.9 ± 1.1	0491A	D–790
Flexural modulus (psi × 10^6)	1.2	0.7	1.2	1.5	1.5	0491A	D–790
Compressive strength (psi)	14,000	10,000	14,000	18,000	18,000	0533	D–695
Tensile strength (psi)	5,000	4,500	5,200	5,000	5,000	—	D–638
Izod impact strength							
(ft/lb/in. notch)[c]	0.34	0.34	0.26	0.24	0.24	0498	D–256
Water absorption (%)	0.1	0.1	0.1	0.1	0.1	0248	D–570
Flammability, U.L. 94, 0.0625 in.[d]	V-0	V-0	V-0	V-0	V-0	—	—
Thermal conductivity							
(cal/sec °C-cm 10^{-4}	15	14	16	18	18	colora	—
Thermal expansion,[e]							
Measured parallel to flow							
50-150°C, in./in.°C 10^{-6}	22	22	24	25	25	0563	—
150–250°C, in./in.°C 10^{-6}	27	29	28	26	28	0563	—
Measured perpendicular to flow							
50–150°C, in./in.°C × 10^{-6}	40	43	38	34	35	0563	—
150–250°C, in./in.°C × 10^{-6}	46	52	39	35	38	0563	—
Stability to 70°C/93% RH							
after 10,000 hr,							
weight change (%)	+ 1.4	+ 1.5	—	—	—	—	—
Dimension change (%)	+ 0.14	+ 0.16	—	—	—	—	—

Electrical							
Arc resistance, seconds	290	270	280	300	—	0171	D–495
Dielectric strength, in oil, V/MIl[f]	320	310	290	300	370	0114	D–149
Volume resistivity							
Condition A[g] Ω-cm × 10^{15}	2.8	2.7	2.8	2.8	2.6	0249	D–257
Condition D[h] Ω-cm × 10^{15}	2.7	2.7	2.8	2.8	2.5	0249	D–257
Dielectric constant at 10^6 Hz							
Condition A[g]	3.55	3.61	3.68	3.60	3.49	0543	D–150
Condition D[h]	3.60	3.64	3.72	3.65	3.53	0543	D–150
Dissipation factor at 10^6 Hz							
Condition A[g]	0.0016	0.0018	0.0018	0.0014	0.0012	0543	D–150
Condition D[h]	0.0019	0.0027	0.0019	0.0019	0.0015	0543	D–150

[a]All values shown are determined periodically and must be considered typical. Should a property be judged to be pertinent in a given application, verification must be made by the user prior to use.
[b]Dow Corning Test Method. Similar to ASTM method shown. Copies available upon request.
[c]Not corrected for energy to toss.
[d]UL yellow card listed.
[e]Test specimens 1.5″ × 0.25″ × 0.10″ Transfer molded bars post cured 2 hours at 200°C.
[f]Tested under 1/4 inch electrodes, 500 volts/second rise, 1/8 inch-thick specimen.
[g]Condition A = As received.
[h]Condition D = After 24-hour immersion in distilled water at 23°C (75°F).

Reinforcements

- Quartz
- "E" type glass fibers
- Fused silica

Typical Data Sheet Values

Typical values are given in Table 8.[9]

Applications

- Encapsulation of semiconductor devices

  Microcircuits
  Capacitors
  Resistors

- Electrical connectors
- Seals
- Gaskets
- O-rings
- Terminal and plug covers

Suppliers U.S.A.

Dow Corning Corp., Midland, MI 48686-0994
ICI/Fiberite, 501 W. 3rd St., Winona, MN 55987
General Electric Co., Silicone Prod. Div., Waterford, NY 12188
Thermoset Plastics Inc., Indianapolis, IN 46220

1.7 COMPOSITES (THERMOSETTING)

A composite material could be defined as two or more substances that, when combined, will retain their individual identities but do interface chemically and physically when subjected to specific conditions of either time, temperature, pressure, or chemical catalyzation.

Composites are available with different resin systems and reinforcement combinations designed to meet individual applicational requirements. They are produced in several physical forms and can be shaped with conventional molding techniques as well as a variety of other less conventional methods such as resin transfer molding (RTM), resin injection molding (RIM) structural resin injection molding (SRIM), hand lay-up, and autoclave molding (Vacuum Bag) to mention a few.

RESINS

Polyesters

Unsaturated polyesters, derived from maleic anhydride or fumeric acid, when dissolved in reactive toluene monomer (styrene/vinyl toluene/diallyl phthalate/methyl methacrylate) will produce a solution with a typical viscosity range of 200–2000 cP. With the addition of heat and a free radical initiator (organic peroxide) a cross-linking reaction takes place, resulting in the conversion of the low viscosity solution into a three-dimensional thermoset plastic. Since unsaturated polyesters cure without volatile by-products, they can be cured under little or no pressure, making hand lay-ups possible.

Room temperature reaction can also take place with the use of peroxides and a suitable activator.

There are five widely used polyester resins available:

1. Ortho: Most widely used but has some definite limitations in both chemical and thermal properties.
2. Iso: Better mechanical and chemical properties than the ortho but higher in cost.
3. Bisphenol A (BPA) fumarates: Very good chemical and thermal properties—considered for high performance.
4. Chlorendics: Excellent chemical properties combined with some flame retardancy.
5. Vinyl esters: Exceptional mechanical and chemical properties but higher cost.

Polyimides (Thermosetting)

Polyimides are produced as two types:

1. Condensation (basically thermoplastic): Derived from polyamic acids by either chemical or thermal treatment over a temperature range from room temperature to 570°F (300°C).
2. Addition (basically thermoset): Bismaleimides. Conversion of bismaleimide to polymeric systems requires copolymerization with amines or nucleophylic monomers via a Michael addition reaction, or with olefinic or acetylenic monomers.

The thermosetting polyimides have temperature performance capabilities in the 316°F (600°C) range.

Epoxies

The most common basic process for producing epoxy resins is the reaction of epichlorohydrin with compounds containing reactive hydrogen atoms, such as phenols or amines. These resins find extensive use due to their mechanical, physical, chemical, and thermal properties along with their ease of processing.

Polyureas

This relatively new resin system involves a combination of novel MDI prepolymers and either amine or imino-functional polyether polyols. Such resin systems can be reinforced with milled glass fibers, flaked glass, Wollastonite, or treated mica, depending on the compound requirements as to processability or final surface finish.

REINFORCEMENT SYSTEMS

The employment of various reinforcing materials in a wide variety of forms into the composite combination or matrix greatly enhances the property profile of the composite product with strength being the primary benefit. However, property enhancement is not confined to mechanical and physical values. Thermal and chemical resistance and light weight are added property benefits that are attractive to the product designer.

Some of the basic reinforcement materials are

- Glass fibers

  Chopped rovings
  Continuous lengths
  Discontinuous lengths
  Woven mats and three-dimensional woven shapes
  Filament windings

- Carbon fibers
- Organic fibers
- Boron fibers
- Continuous filament yarns
- High alumina fibers
- Rovings
- Woven fabrics
- Wollastonite
- Treated mica
- Organic fibers

Continuous filament yarns
Rovings
Spunlaced sheets
Woven fabrics

PROCESSING METHODS

The conversion of the composite material into a final cured state of the proper design configuration is accomplished in several different ways, usually dependent on

- Composite material

  Type of resin and reinforcement

- Part configuration
- Part requirements

  Mechanical
  Chemical
  Electrical
  Thermal
  Aesthetics—colorability—surface finish

Composite materials can be shaped, formed, or molded in the conventional thermosetting molding processes such as compression, transfer, and injection—again depending on the factors listed above. In addition there are some other techniques employed that better preserve the integrity of the reinforcement system, producing parts with exceptional strength characteristics:

Resin Transfer Molding (RTM)

RTM is accomplished with the use of a reinforcement system that has been preshaped to fit the mold cavity and a low-pressure resin—often liquid—that is injected into a closed mold. The injection process can be assisted with the use of a vacuum system to provide for evacuation of the volatiles and to gain molded density.

Reaction Injection Molding (RIM)

RIM molding employs a closed mold into which is force fed a combination of two resin systems that have been premixed in an impingement chamber. The two resins react chemically within the mold cavity in a thermosetting type reaction achieving a cured state.

Structural Reaction Injection Molding (SRIM)

SRIM utilizes a preshaped reinforcement system preloaded into the cavity, and, with the mold in the closed position, the resin mix from the impingement chamber is injected into the mold where the curing reaction takes place.

Hand Lay-up

This is an open mold technique that has been used for the production of a wide variety of articles where the resin and reinforcement systems can be tailored to meet specific end use requirements of applications such as skis, boat hulls, tubs, and shower stalls.

The mold cavity is sprayed with a gel coat to a desired thickness to produce a desired surface finish or color. The reinforcement (usually glass fibers, mats, or sheets) is then hand rolled to secure proper wetting of the fibers as well as a desired wall thickness. Final curing takes place through the catalyst or hardener reaction.

Matched metal molds are often used with hand lay-up techniques to achieve better surface finish on both sides of the molded article.

APPLICATIONS

- Epoxy resin-based composites

  Printed circuit boards
  Automotive components
  Aerospace/satellite components
  Pressure vessels

- Polyimide resin-based composites

  Automotive components
  Aerospace/aircraft components

- Polyester resin-based composites

  Marine equipment
  Residential components and appliances
  Recreational equipment
  Business machine housings

- Polyurea resin-based composites

  Automotive

  Exterior body panels
  Bumpers
  Tail gates
  Truck boxes

Table 9 provides a look at the composite market for 1989 and a forecast for 1990 that gives a fair idea of the future of this thermoset product.

Table 9. SPI 1990 Forecast for Composites: Projected Shipments by Market Sector (millions of lb)

Market	1989	1990	% change 1990/1989	Market share (%)
Aircraft/aerospace/military	41.4	42.0	2.2	1.6
Appliance/business equipment	151.5	154.9	2.2	5.9
Construction	478.8	482.0	0.7	18.3
Consumer products	162.7	164.4	1.0	6.2
Corrosion-resistant equipment	349.0	354.5	1.6	13.5
Electrical/electronics	231.0	236.0	2.2	9.0
Marine	412.1	414.0	0.5	15.8
Transportation	684.8	701.5	2.4	26.7
Other	78.0	80.0	2.6	3.0
Total	2,589.3	2,629.3	1.5	100.0

Source: SPI Composites Institute. Figures include reinforced thermoset and thermoplastic composites, reinforcements, and fillers.

RECOMMENDED READING

Engineered Materials Handbook COMPOSITES, Volume I. American Society for Metals, International Metals Park, OH 44073.

Engineers' Guide to Composite Materials. American Society for Metals, Metals Park, OH 44073.

REFERENCES

1. Rogers Corp. MMD, Manchester, CT.
2. *Textbook of Polymer Science*, F. W. Billmeyer, Jr.
3. American Cyanamide Co., Wayne, NJ.

4. Plastics Engineering Co., Sheboygan, WI.
5. Rogers Corp. MMD, Manchester, CT.
6. *Textbook of Polymer Science*, F. W. Billmeyer, Jr.
7. Rogers Corp. MMD, Manchester, CT.
8. Rogers Corp. MMD, Manchester, CT.
9. DOW Corning Corp., Midland, MI.

CHAPTER 2

PRODUCTION METHODS AND EQUIPMENT

The choice of methods and equipment for use in the production of thermosetting molding compounds is, to a great extent, determined by the type of "filler" or "reinforcement" to be used in the products coming off the production lines.

The necessity for retaining the integrity of the reinforcement in order to meet certain mechanical values such as impact and flexural strength dictates methods and equipment that would provide the least amount of reinforcement degradation throughout the entire compounding process. Such concern regarding degradation was considerably less for those products using fillers/fibers such as wood flour, short glass fibers, free flowing minerals, or short fiber cotton and acrylic fibers. Generally, products utilizing those fillers exhibit impact values of 0.3–1.0 ft-lb and flexural values of 8,000–14,000 psi whereas the impact values of the bulkier, higher strength products are in the range of 1.0–30.0 ft-lb and flexural values in the range of 16,000–30,000 psi. These higher values could best be attained with the use of reinforcements such as long fiber glass, cotton, minerals, or organic materials, and it is imperative that these reinforcements not suffer any degrading effects from the mixing and densification process. A process system involving the use of solvents, coupled with mixing equipment capable of imposing a minimum of stress on the reinforcement, meets these demands. There are variations of both the process and the equipment that are related to the specific resin sytem and color requirements, but the basic premise of retention of reinforcement integrity is predominant in all cases.

2.1 WET PROCESS

The higher strength products require the use of the solvent, or "wet," system whereas the lower strength products can be made with either the wet or dry process. The following will describe the three wet processes generally employed:

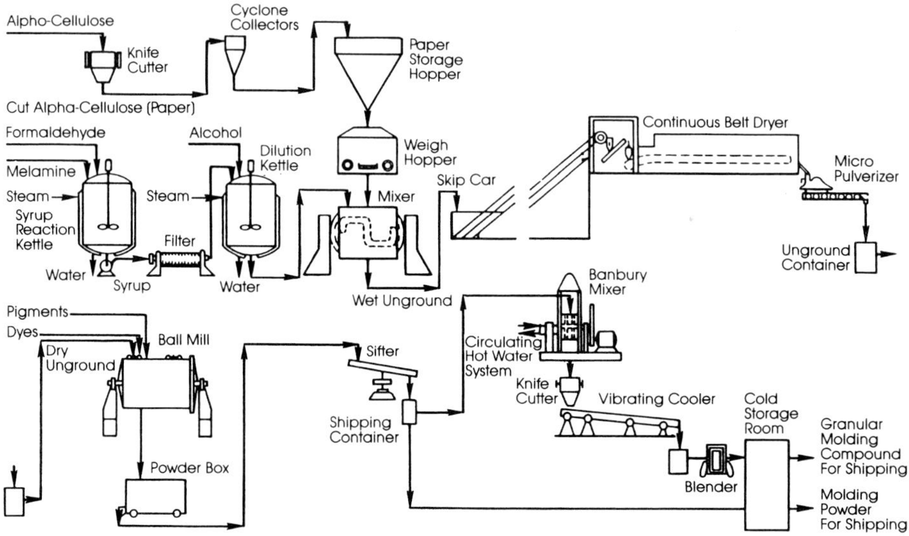

Figure 1. Schematic of steps in preparation of alpha cellulose filled amino molding compounds.

2.1.1 Granular (Free Flowing) Products (Figure 1)[1]

Equipment:

Mixers
Roll mill
Calendering mill
Granulator
Screens
Conical blenders

Generally, the reinforcement, lubricant, and pigment are thoroughly dry mixed without heat and then a solution of liquid resin with catalyst is added, the lid closed, and a vacuum is drawn as the mix is completed. The resultant dough-like mass is conveyed to the roll mill where heat, pressure, and time are applied for the purpose of attaining a proper degree of homogeneity, density, temperature, and preliminary reaction. Control points are the roll mill temperature and speed, the "nip" size (the distance between the rolls that creates the needed pressure), and the time needed to achieve the desired condition. With the rolls operating in opposite directions, the compound can be made to adhere to one of the rolls as a relatively thin sheet of uniform density and homogeneity.

This thin sheet of heated compound is stripped from the mill and fed onto the unheated calendering mill, where it becomes a sheet of specific thickness

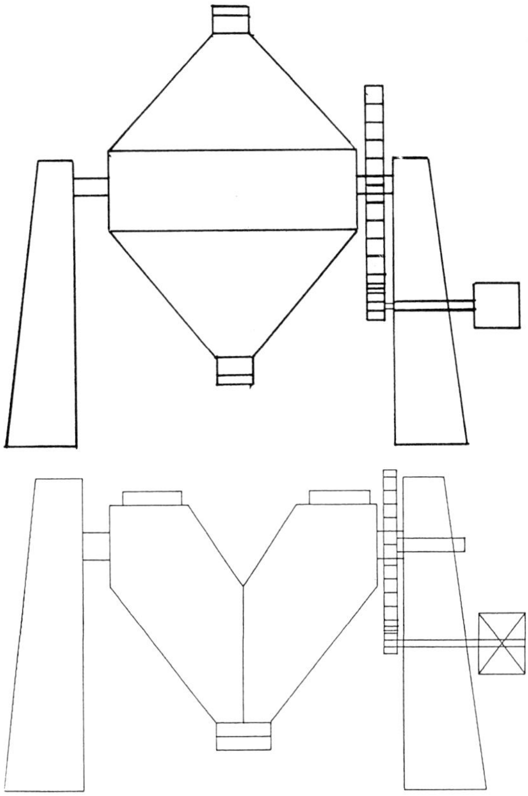

Figure 2. (A) Single cone blender. (B) Twin cone blender.

and width with its preliminary reaction arrested by the lowered temperature. The sheet is then transported on a conveyor belt operating at a set speed, under a cooling area prior to its introduction to the granulator. The sheet width, created by the calendering mill, is determined by the size of the granulator opening. Temperature and thickness are important because if the sheet is too warm it will not cut cleanly, fouling up the blades, whereas a sheet that is too cool will shatter into many large pieces. A sheet too thick will generate oversize particles, whereas a sheet too thin will generate excess dust or "fines."

A fairly uniform particle size from the properly sized and cooled sheet will then be conveyed to a deck of screens that will determine its final particle size distribution. Oversize particles go back to the granulator with the dust retained for use in future batches.

The final granulation will often be specified by the molder and will depend on the type of molding process being employed. Automatic compression molding, utilizing a volumetric feed, will require an extremely uniform granulation for controlling the shot weight as well as having excellent pourability. Automatic preforming operations requiring precise preform weights and automatic transfer or injection molding methods, and also requiring precision feeding, will utilize this same automatic molding granulation (AMG). Molding operations where such precision is not needed will utilize a somewhat coarser particle size with a dust content sufficient to ensure good preform density.

Mixer batch sizes will be in the 200 lb range with each batch identified by product, color, granulation, and flow as it comes off the screens. These batches are then poured into conical blenders (Figure 2) for the purpose of achieving a final blend conforming to a specific set of standards—color, granulation, and flow being the most important.

2.1.2 Pelletized (Free Flowing) Products

Equipment:

Kneader/mixer
Extruder
Dryer
Blender

This type of wet process (Figure 3) is capable of producing products with impact values up to 2.0 ft-lb using water as the solvent. The ingredients are thoroughly mixed in the kneader and then auger fed to the extruder, which is

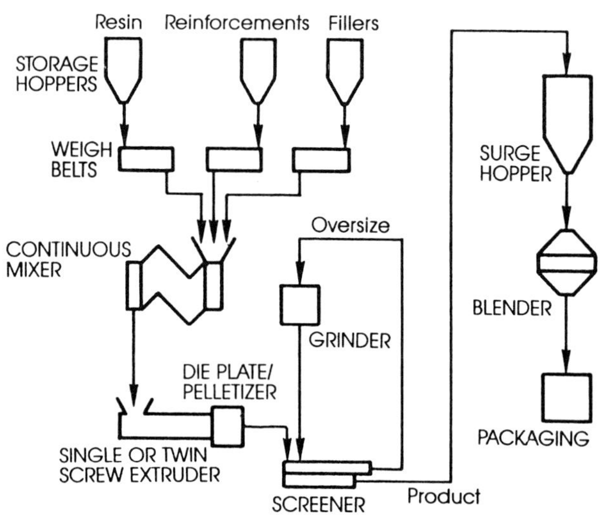

Figure 3. Continuous production method.

heated to soften the mass as well as to densify it prior to being forced by the screw through a set of holes in a plate at the head of the extruder. The holes determine the diameter of the extrudate and the fly-cutters on the outside face of the extruder head determine the length. These extruded pellets, still somewhat damp, are conveyed to long cylindrical heater drums for drying. As they exit from the dryer they are collected in drums for further blending. The blend sizes are in the 2,000–10,000 lb range and are identified as to product, color, granulation, and flow.

2.1.3 High Strength (Nonpouring) Products

Equipment:

Mixers
Mobile carts
Air drying room
Prebreaker
Hammer mill
Blenders
Extruders

The process and equipment to produce these compounds (Figure 4) are designed to provide minimum reinforcement degradation with the end product capable of meeting mechanical strength requirements usually set by either military or commercial specifications.

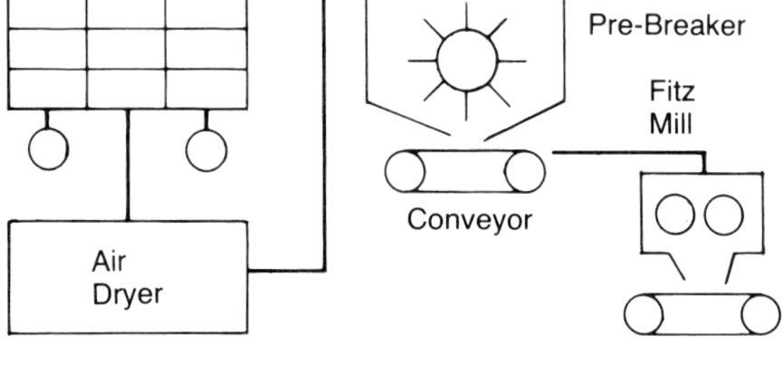

Figure 4. High-strength production method.

The initial mixing procedure entails the use of a solvent into which all of the ingredients except the reinforcement are thoroughly dispersed and mixed. Once the mix has been completed the reinforcement is added with the mixing time kept to a minimum to preserve the reinforcement integrity. Solvent recovery is accomplished during this phase of the operation as well as during the subsequent air drying cycle that occurs next.

The wet mass of mixed ingredients is removed from the mixer and spread onto individual wire mesh trays about 3 ft long by 2 ft wide and some 2–3 in. in depth, capable of holding some 25 lbs each. Approximately 20 trays are then loaded into a mobile trolley cart and delivered to the drying room, where they remain until the formulation is completely dry. From the drying room the trolley goes to the prebreaker—a rotating steel shaft of 3-in. nominal diameter containing a series of hardened steel "fingers" that tear apart the rigid slab coming out of the tray and feeding the large particles to a hammer mill for further particle reduction. The final particle size is controlled by the openings in the bottom screen.

In this process there is no need for blending to achieve a specific flow because the flow properties of the compound are the result of its resin, catalyst, and reinforcement ratio and the batches are very uniform in this respect.

The final compound coming out of the hammer mill is conveyed directly to a storage unit from which the finished product can be poured into 75–150 lb cartons for shipment. Each carton is identified as to the product, color, and batch number.

There is another version of the wet process used in the manufacture of high-strength thermoset polyester compounds in which the doughy mass from the mixers is fed into twin screw extruders to be formed into round "logs" or "ropes" as they exit the nozzle end of the extruder. These compounds are mostly glass filled and supplied in varying lengths and diameters for further size reduction by molders as fits their needs.

2.2 DRY PROCESS

2.2.1 Batch/Blend Method

Equipment:

Ribbon or conical mixers
Auger feeders
Roll mill
Calender mill
Cooling chamber
Granulator screens
Blenders

The dry process (Figure 5) is quite similiar to that of the wet process except that the initial mix is accomplished with dry ingredients and then fed by augers

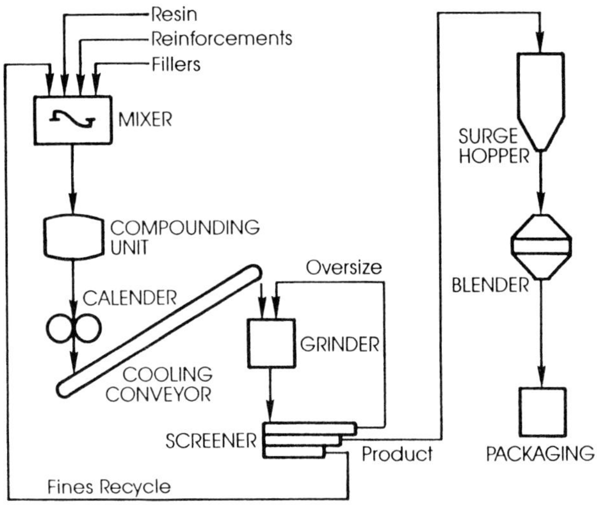

Figure 5. Batch/blend production method.

to the heated roll mill for the compounding process. Once the proper sheet density, temperature, and homogeneity are reached the compound is fed into the calendering mill for width and thickness sizing. From the calender mill the sheet goes through the cooling chamber, into the granulator, and then onto a deck of screens for particle size and distribution. Again, the batch sizes will run in the 200 lb range with each batch carefully segregated as to its flow so that the final flow target can be attained through the blending of several batches into a final blend.

2.2.2 High-Volume Production

There are a number of thermosetting applications calling for very large quantities of a single product or a variety of very similiar products that would lend themselves to more automated production lines utilizing twin screw extruders or "compounders" into which the ingredients are continuously fed to the mixing head by which the mass is "worked" by the speed and back pressure on the screw to simulate the action of the heated roll mill (Figure 6).[2,3]

The densified, homogenized mass travels the length of the extruder barrel and is forced out through the head either in sheet form for granulating or through holes in the head plate, resulting in round extrudates that are cut off in specific lengths. Production from such equipment is very high and cleanup time when changing from one product or color to another is relatively short due to the equipment design. All working components are easily exposed during shutdown, or it is possible to "purge" the entire mixing area without dismantling the extruder. The process is primarily used for the high-volume production of a single product or color.

A

B

Figure 6. (A) Open Buss kneader, with external screw bearing for thermosets compounding. The external bearing is used for pastel colors in urea and melamine compounds. Epoxies, polyesters with and without glass fiber, and phenolics, the external bearing is not required. (B) Werner Pfleiderer extruder compounder. (C) A schematic of the operating principle of the Werner Pfleiderer compounder.

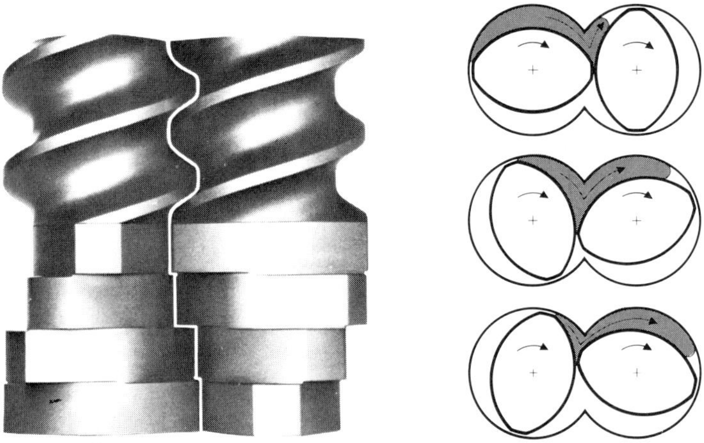

C Self-cleaning profile Material transport in the WPC extruder

2.3 SHEET PROCESS (COMPOSITE COMPOUNDS)

2.3.1 SMC Products (Sheet Molding Compound)

Thermosetting sheet molding compounds are produced with one of several different resin systems—phenolic, epoxy, or thermoset polyesters—to meet the varying needs of the marketplace. The manufacturing processes for each system are quite similiar, all involving the production of continuous lengths of the compound, with the doughy compound and its reinforcement being covered on both upper and lower surfaces with a thin film of polyethylene (Figure 7).

The finished product is conveyed to a mandrel with the sheet rolled onto the mandrel until it arrives at a preset weight and is then cut off. The sheets are generally 0.075–0.250 in. thick, 4 ft wide, with weight varying with the length—usually about 50 ft.

Surface quality of the molded parts will be determined by compound chemistry, process controls, molding parameters, molding press platen parallelism and finish, and, finally, the priming/paint system in use.

The early SMC formulations were made up of 20–30% chopped glass rovings, unsaturated polyeaster resin, 40–50% fine-particle-size calcium carbonate filler, catalyst, pigment, and other modifiers. Today, the formulations contain high fiber loadings using E-glass, S-2 glass, or carbon (PAN) fiber reinforcements. The fiber reinforcement can be either chopped or in continuous lengths. The resin system will be epoxy, polyester, or vinyl ester to fit the special needs of the marketplace.

SMC compounds can be compression, transfer, or even injection molded if the injection equipment is augmented by special feeding devices.

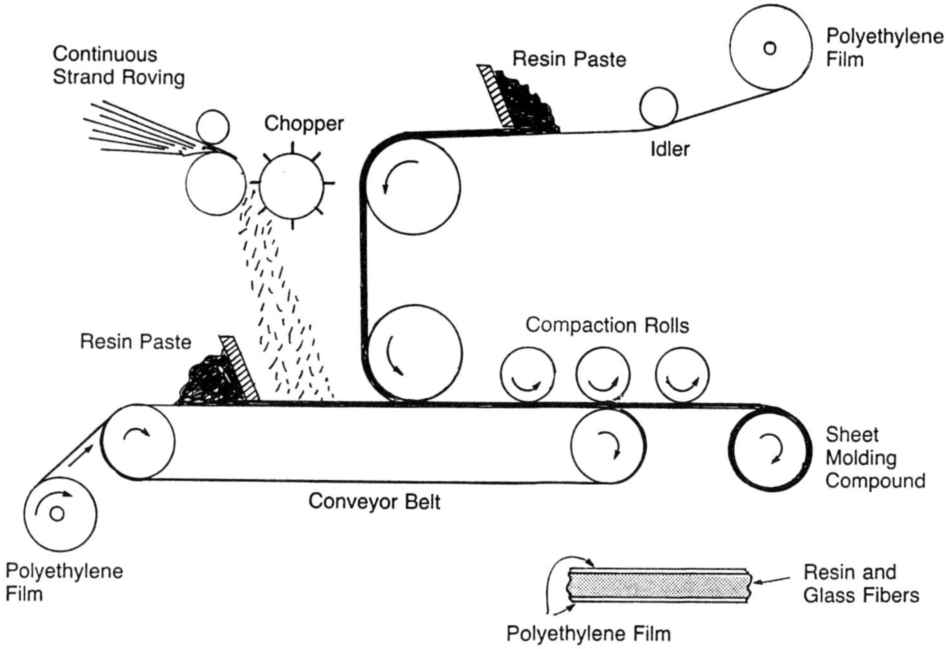

Figure 7. Sheet molding process.

2.3.2 Thick Molding Compounds (TMC)

Thick molding compound (Figure 8)[4] differs from the SMC product in that the glass fibers are wetted between the impregnating rollers prior to being deposited onto the moving film. Whereas the SMC sheets are in the 1/4 in. thickness range, the TMC compounds can be produced in thicknesses of up to 2 in. with glass lengths of 1 in. minimum at loading levels of 20–30%.

Generally, the TMC compounds are compression molded using matched metal hardened steel molds. Packaging and shipment of the unmolded TMC are similiar to those used for SMC compounds.

2.3.3 BMC Products (Bulk Molding Compounds)

Bulk molding compounds are produced a batch mix technique in which the resin, lubricants, catalysts, and chopped glass fibers (1/8–1/2 in. long) are compounded in relatively low intensity mixers—the intent being to achieve homogeneity with the least degradation of the glass fibers, which are at a 20–30% loading. The product is available in bulk form shipped in vapor barriered cartons with the compound inside a sealed polyethylene bag. The compound can

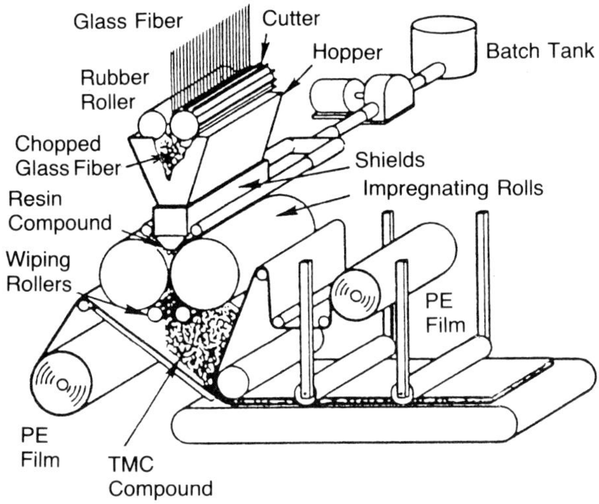

Figure 8. Schematic of the TMC process.

also be supplied in "rope" or "log" form in any length and diameter specified by the customer using an extruder to shape the final product.

The BMC products are compression, transfer, and injection molded. The injection machines have been especially designed to provide accurate feeding of the bulky material to the screw or to the transfer ram, depending on whether the injection machine is a screw or a plunger type.

2.4 QUALITY ASSURANCE

Regardless of the process or equipment in use, a comprehensive system for ensuring both the supplier and the customer that the product coming off the production line does meet the required standards of quality is requisite. Such a system has its origin in the compounder's formulation and processing specifications.

The formulation is designed to meet certain standards relating to a profile of the following property values:

- Flow properties
- Electrical (insulation)
- Mechanical (strength)
- Chemical resistance
- Thermal resistance
- Flame resistance
- Surface finish

- Color
- Weather resistance

The standards for most thermosets are derived from the three major specifications referred to in Chapter 1:

- Mil-M-14
- ASTM
- U.L.

With a well-defined set of standards to meet, the designer's task becomes one of selecting individual ingredients that, within a cost range, will not only create a compound meeting those standards but will also function properly in the compounding process. Along with the formulation must come a set of processing procedures governing the process and spelling out the various checkpoints along the line that will aid in controlling the process.

The formulation provides the information relating to the acquisition of the raw materials and the basic standards by which these materials will be accepted for incorporation into the original mix. Each of the following ingredients contributes its own characteristics to the compound's successful manufacture:

- Resin
- Reinforcements/fillers
- Pigments/dyes
- Catalysts
- Lubricants (internal/external)
- Solvents

The specific characteristics of each of these ingredients are controllable and measureable with testing equipment and procedures at either the supplier's or the compounder's facility. Very often the supplier will furnish test data relating to the product on which the compounder can rely on without retaining his own elaborate testing equipment.

The following is a breakdown, by ingredient, of the vital characteristics to be closely monitored both before the initial mix has started and then during the production procedure:

Resin

- Viscosity
- Gel time
- Cure rate
- Solubility

Reinforcements/fillers

- Aspect ratio
- Moisture content
- Fiber size
- Purity
- Color

Pigments/dyes

- Solubility
- Coatability
- Thermal stability

Catalysts

- Solubility
- Reaction temperature
- Purity

Lubricants

- Solubility
- Melting point
- Purity

Solvents

- Purity
- Toxicity

Processing procedures and checkpoints for process control during a production run are generally determined by the effects of each operation on the ultimate quality of the compound as well as on the ability of the compound to meet its specifications. The electrical, mechanical, chemical, and thermal properties are derived primarily from the ingredient mix whereas the flow (rheology) properties—*very important to the molder*—can be seriously affected by the compounding procedures, where temperature and time come into play. Some of the flow properties result from the resin/catalyst/reinforcement ratios, but the application of heat and time during the compounding must be carefully monitored and controlled to be able to meet the rheological targets consistently.

Flow checks should be made as soon as practicable, and this check will also serve to detect color or precure contamination. Once the compound has been granulated and screened to a specific particle size and distribution, a screen analysis can be run along with batch flow measurements needed to determine the final blending procedures.

Compounds manufactured to meet Mil specifications are subjected to a certification process that involves the documentation of actual property values derived from testing the compound for the following:

- Specific gravity
- Shrinkage
- Arc resistance
- Dielectric constant
- Dielectric strength
- Impact strength
- Flexural strength
- Volume and surface resistivity

Certification will cover a blend (2,500 lb or more), and the actual values will be furnished to the customer on request. An alternate certification is for the compounder to supply the customer with a letter of compliance in which the compounder responsibly states that the blend involved meets all of the requirements of the specification.

The following is a list of testing equipment employed by compound manufacturers to control and monitor their production, in ongoing R & D programs, and for troubleshooting field problems:

- Unitron metallograph: A sophisticated metal detection device used to check production as well as in troubleshooting areas.
- Scanning electron microscope (SEM): SEM is employed to examine surfaces from low magnification to 100,000× enlargement. An excellent research and problem solving tool.
- Dynamic mechanical spectrometer (DMS): This instrument measures the viscoelastic properties of polymers, determining a compound's viscosity and elastic modulus, following the change of these properties over time and changes in temperature.
- Infrared spectrophotometer: Essentially a chemist's tool, the infrared spectrophotometer provides information concerning a compound's composition. It helps the researcher understand the degradation of compounds; find replacements for resins, fillers, and reinforcing agents; and evaluate the purity of resins, fillers, catalysts, reinforcements, and lubricants.
- Capillary rheometer: The Monsanto capillary rheometer measures viscosity properties of polymers, providing a direct measure of viscosity and the change of viscosity with time and flow rate at plastication temperatures. The capillary orifice simulates the gate and runner system of actual molding conditions, thus providing valuable flow information of molding compounds.
- Brabender plasticorder: The plasticorder is a small mixer capable of measuring the viscosity and the gel time of thermosetting compounds. The results can be correlated to a compound's performance.

- High-pressure liquid chromatograph (HPLC): This machine can have two modes:

  1. Gel permeation chromatograph—measuring the molecular weight distribution and average molecular weight of the molecules in a sample of a compound.
  2. Liquid chromatograph—detecting and measuring the amount of chemical constituents present in a compound.

- Thermal analysis equipment: Du Pont's thermal analysis equipment has four modules to provide information regarding the effect of temperature on a compound's physical properties:

  1. Differential scanning calorimeter (DSC): DSC measures heat uptake or heat release of a compound as the temperature is raised and also measures heat effects associated with material transitions, such as melting.
  2. Thermal analyzer (TMA): This module measures the variation in the length of a sample as temperature is increased. Results are helpful in matching a compound's thermal expansion with that of another compound or a metal. TMA measures thermal transition points such as predicting the point and rate at which a compound will melt as well as determining the temperature at which blistering will occur if a molded part has not been properly postbaked.
  3. Dynamic mechanical analyzer (DMA): DMA measures a compound's modulus (stiffness) as its temperature is raised. Testing done with this instrument has provided interesting insights into the properties of phenolics as well as those of diallyl phthalates, thermoset polyesters, and thermoplastic compounds, indicating the ability of thermosets to retain their modulus at elevated temperatures; thermoplastic compounds, in general, do not have this capability.
  4. Thermogravimetric analyzer (TGA): TGA measures weight changes in a sample as temperature is varied, providing a useful means of determining degradative processes and heat resistance in polymeric compounds.

- Particle size analyzer (PSA): PSA is an accurate and automatic development tool that allows for a very rapid measurement of particle size distribution in powder or slurry compounds.
- Humidity chambers: These are employed to measure the effects of temperature/humidity cycles on molded parts.
- Instron testers: Instron testers will measure flexural, tensile and compressive strength as well as stress/strain curves at ambient temperatures, and when fitted with an environmental chamber the flexural and tensile tests can run at elevated temperatures.

2.5 RHEOLOGY—FLOW PROPERTIES and TESTING

One of the most important characteristics of thermosetting molding compounds centers around the rheology or flow of these materials when they are subjected to molding conditions of applied heat, pressure, and time. Because they undergo a dynamic chemical reaction as a result of those three conditions, their flow properties are greatly affected and are limited in duration.

As a general rule, the speed of the reaction will double for every 10°C increase in temperature. The typical thermosetting reaction can be best described by Figure 9.

At the top left side (A) the compound is at room temperature and at zero (0) pressure. With pressure applied and the compound increasing in temperature, its viscosity decreases as shown along the slope at (B). This decrease in viscosity continues, reaching a peak of flow just prior to the very rapid acceleration of the reaction depicted at (C) and (D), with complete cure or reaction cessation at (D).

Each individual compound will possess its own flow characteristics, which will be affected by the following:

- Resin/catalyst/reinforcement ratio
- Type and content of catalyst
- Type and content of reinforcement
- Type and content of resin
- Type and content of internal lubricant

The compound designer will have a "flow profile" target that will depend, when the formulation is put together, on several factors:

- Molding method:

  Compression
  Transfer
  Injection

- Flow specifications:

  Stiff
  Medium
  Soft
  Cure rate/flow time

The flow characteristics of any thermosetting molding compound are measureable and controllable with several procedures available for this purpose, all of which will measure one or more of the following:

- Rate of cure
- Speed of flow

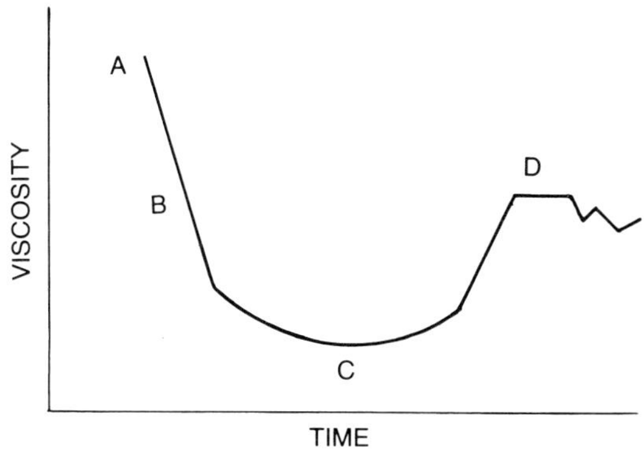

Figure 9. Typical thermosetting reaction.

- Distance of flow
- Amount of compound used during flow time

There are several flow testing procedures employed, with each procedure designed to provide specific information based on one or more molding methods. The flow characteristics for compression molding are, for instance, quite different than those for either transfer or injection molding, and the tests will reflect those differences. The commonly used test procedures are

- Cup closing time (ASTM)
- Disc flow I and II
- Orifice flow
- Spiral flow
- Brabender viscosity

The cup, disc, orifice, and spiral flow tests will keep three elements constant when the tests are conducted:

- Amount of compound—charge weight
- Mold temperature—usually 300°F
- Molding pressure—usually 1,000 psi

The test procedures for each flow testing technique are described as follows:

2.5.1 Compression Molding

2.5.1.1 Cup Closing (Figure 10)

This test requires that a room temperature charge of compound be placed into the lower cavity and the press closed. The time required to completely close the mold is recorded in seconds; the longer the time required, the stiffer the flow.

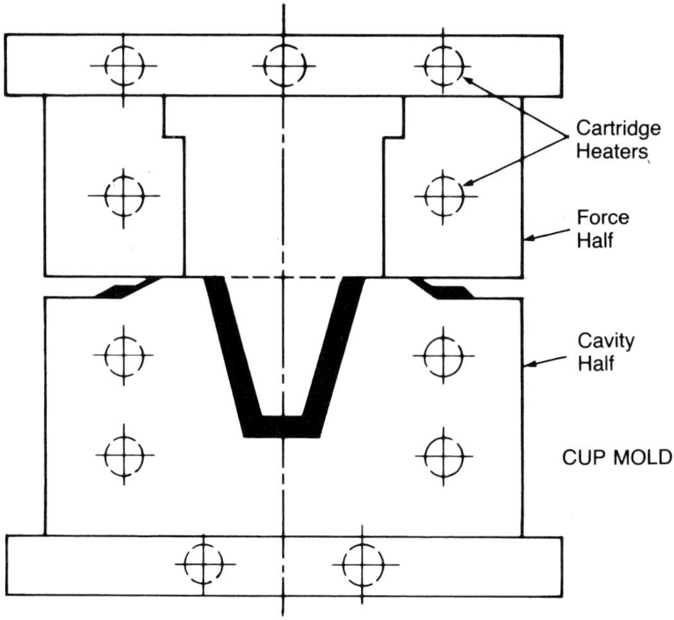

Figure 10. Cup closing test.

Stiff flow compounds will be in the 15 sec and upward range; the medium flow compounds will be in the 8–14 sec range; and the soft flow compounds in the less than 8 sec flow range. With this knowledge available, a molder can specify the flow believed to be most suitable for the particular part to be molded, bearing in mind the overall configuration of the part itself.

The compounder will also use this test as a quality control element that will also show any evidence of precure or contamination.

2.5.1.2 Disc Flow (Figure 11)

This test involves the placing of a measured amount of room temperature compound onto one side of the mold and then closing the mold at the preset pressure, temperature, and closing speed, forming a molded disc. Once thoroughly cured, the disc is removed from the mold and measured, either for disc thickness or diameter.

When considering the disc thickness, the thinner the disc, the softer the flow, and vice versa. Using the diameter as the gauge, the disc is compared with a "target" plot with a series of concentric circles as shown. Here, the larger the diameter, the softer the flow, with the compound being given a "flow number" such as 1S–2S–3S.

This flow test can also be used in conjunction with quality control efforts detecting any precure or contamination.

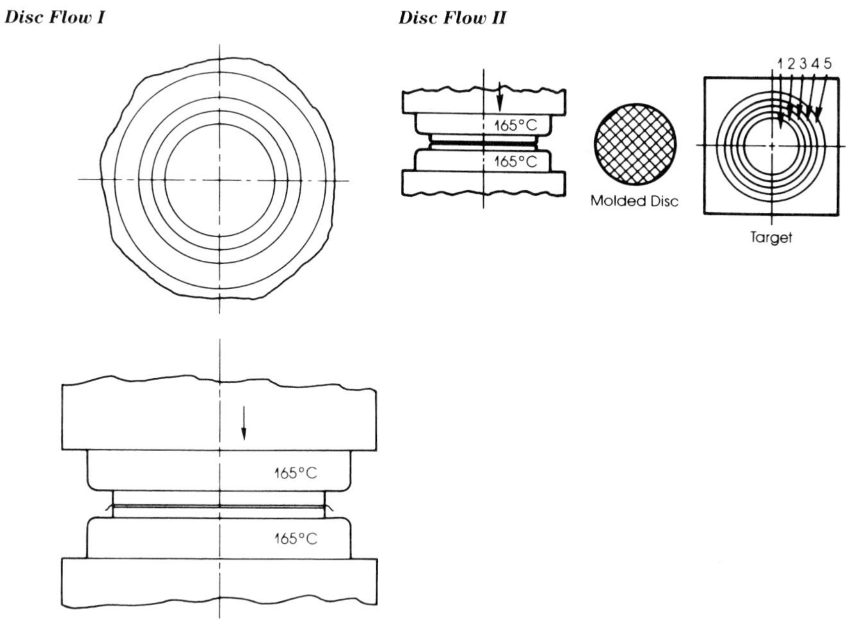

Figure 11. Disc flow test.

2.5.2 Transfer Molding

2.5.2.1 Orifice Flow I (Figure 12)

As shown, the mold is charged with about 12–15 g of room temperature compound and the mold closed under a preset molding pressure. The usual pressures employed will be 600, 900, 1,800, or 2,700 psi depending on the molding procedure for which the compound has been designed. The compound, under pressure, will escape through two small orifices cut into the side of the force plug of the upper mold half. After the mold has been closed completely and the compound thoroughly cured, the material left in the mold is extracted and weighed. The flow is then designated as a percentage of the original amount of material placed in the mold. Also noted in the designation is the pressure employed.

Example: If 30% of the original shot weight is left in the mold and 900 psi was the pressure used, the compound would be designated as 30% at 900 psi.

2.5.2.2 Orifice II (Figure 13)

As illustrated, a measured quantity of room temperature compound is placed in the centrally located "pot" in the middle mold plate and the heated mold is

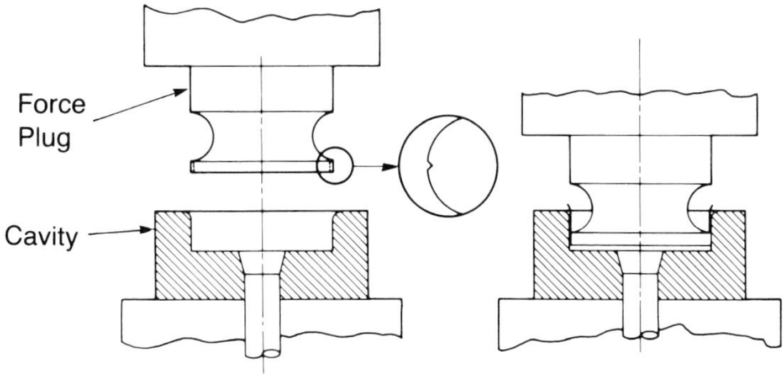

Figure 12. Orifice flow I test.

closed under a preset molding pressure. The quantity is usually 90 grams and the pressure 1000 psi with a mold temperature set at 300°F. As the mold closes the compound is forced out of the pot into a runner that exudes hot compound out the side of the mold. The closing time is recorded from the start of mold closing to the cessation of compound flow, and the exuded compound is then weighed, providing both a time and quantity to the flow property profile. Thus a compound that closed in 30 sec producing a quantity of 60 g would earn a flow designation of 60 g in 30 sec or a flow rate of 2 g/sec. The least material produced in the longest time span would be considered "stiff" whereas a compound with the most material and the shortest time would be designated "soft."

If needed, another factor of flow could be ascertained by cutting off successive sections of the compound as it exits the side of the mold every 20 sec, keeping them segregated so that they can be identified and weighed, providing a time/weight ratio depicting the decrease in viscosity and increase in weight as the compound reacts to the increased temperature.

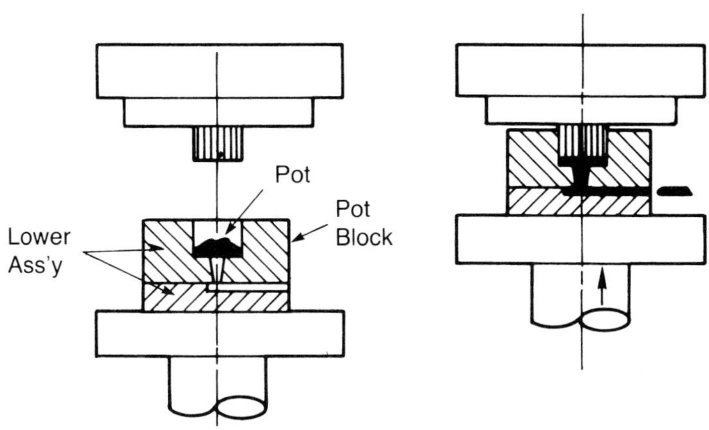

Figure 13. Orifice flow II test.

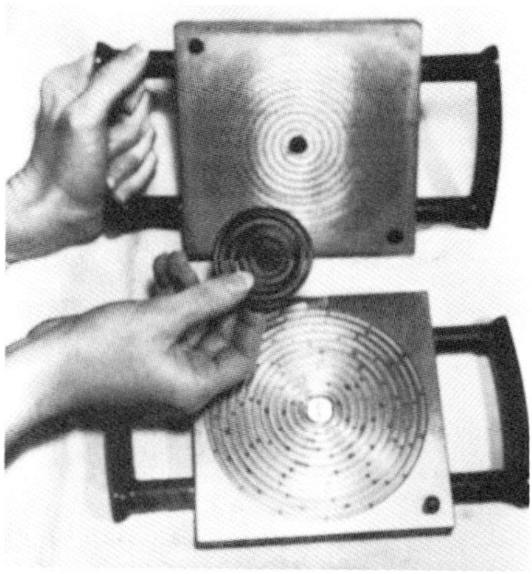

Figure 14. Spiral flow (EMMI) test molds.

2.5.2.3 Spiral Flow (Figures 14 and 15)[5]

As previously mentioned in Chapter 1 in describing epoxy compounds, there are two basic types of epoxy molding compounds—those with flow characteristics similiar to the allyls, aminos, phenolics, and thermoset polyesters and those with extremely soft flow employed in encapsulation molding applications. This has resulted in the formation of two spiral flow test procedures. They both employ a mold using a "pot" into which the compound is placed. The compound is forced out of the pot at a preset pressure and the compound flows into a "runner" system of a specific configuration, and this is where the two procedures primarily differ.

The encapsulation grades of epoxy compounds use a runner configuration of 1/8 in. radius cut in an Archimedes spiral starting from the center of the lower mold half. The runner is marked in linear inches and is 100 in. long. This test is known as the EMMI test and is illustrated in Figure 14.

A similiar test used in the more conventional flow compounds is referred to as the Mesa test, in which the molding procedure is the same but the molding pressure is higher and the runner configuration larger—0.250 wide by 0.033 deep and 50 linear inches long.

In either procedure a designated amount of room temperature compound is fed into the mold and the mold is closed under a designated pressure, temperature, and ram speed with the compound flowing into the runner system until it ceases to flow. The resulting molded runner will reveal its length in linear inches from the settings that have been engraved in linear inches in the bottom of the mold runner. A time recorder can also be utilized to determine the amount of time elapsed in filling out the runner system. In addition, the entire molding can be weighed to determine the quantity of compound used.

Figure 15. Mesa flow test molds.

The flow measurement derived from either system will identify the specific compound's flow by its flow length or a combination of flow length, time, and weight—all very important to the transfer molding process.

2.5.3 Screw Transfer and Screw Injection Molding

2.5.3.1 Brabender Plasticorder

In compression and transfer (plunger) molding the molding compound temperature is controlled with auxiliary equipment not part of the actual molding machine. However, the advent of more sophisticated molding techniques utilizing an extruder for preheating the compound resulted in the entire molding operation becoming more automated with the compound temperature and viscosity being controlled within the molding machine itself. The extruder barrel became a compound reservoir supplying sufficient properly heated compound to mold several "shots" with fresh material being introduced to the barrel from a hopper filled with some 30–50 lb of unheated compound.

This innovation in molding equipment brought with it the need to control the curing reaction and viscosity of the compound when contained within the heated barrel of the extruder.

The Brabender plasticorder, basically a torque rheometer, will measure and record the meter-grams of torque against time while the compound is being mixed, thus producing a flow curve indicating the compound's processability and closely relating to what takes place within either the screw transfer or screw injection process (Figures 16 and 17).[6]

As shown in Figure 16, the device is made up of a heated mixing head using a pair of sigma-shaped roller blades driven by a dynamometer running at a controlled speed. A volumetrically measured amount of room temperature com-

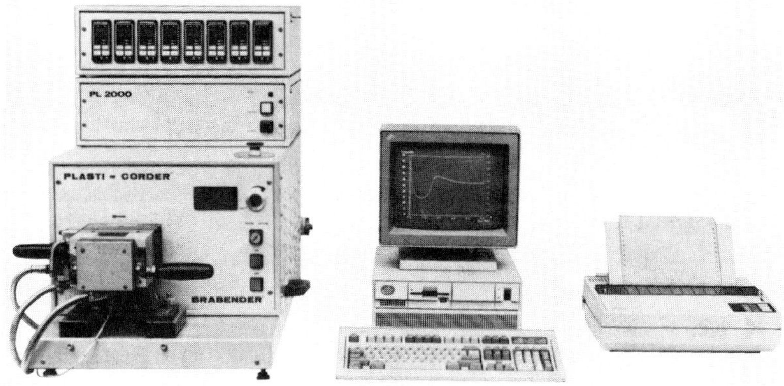

A

B

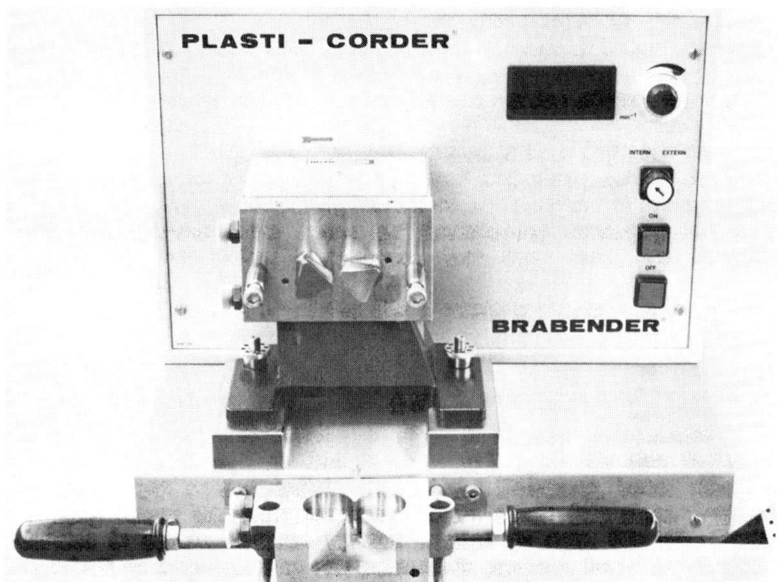

C

Figure 16. Data processing Plasti-Corder. (A) With measuring mixer. (B) With measuring extruder. (C) Front view with opening measuring mixer.

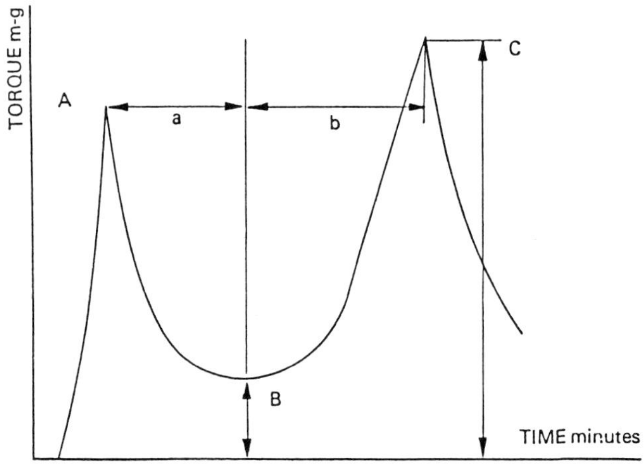

A = Loading torque. End of the filling of the kneader, start of fusion.
B = Minimum viscosity.
C = Cure torque. Maximum torque at grinding moulding compound during curing process.
a = Fusion time.
b = Reaction time, cure time.

A

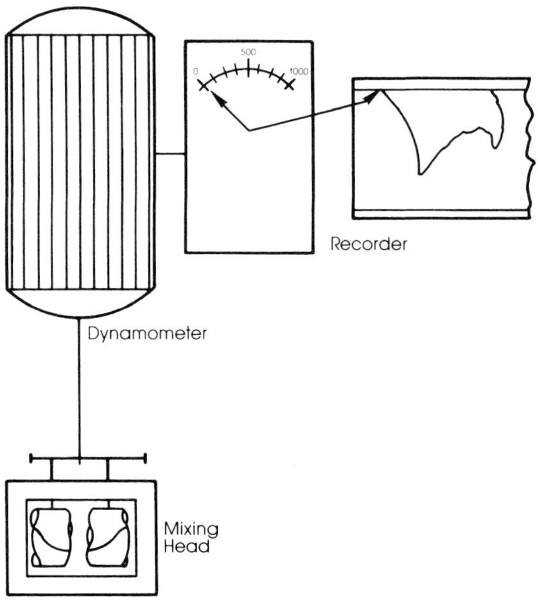

B

Figure 17. Brabender Plastograph Test. (A) Typical Thermoset Curve. (B) Brabender Plasti-Corder Unit. (C) Data sheet from unit with mixer.

BRABENDER

Data-Processing Plasti-Corder PL2000 and Mixer Measuring Head
Flow-Curing Behav. of Crosslink. Polymers acc. to DIN suggestion

Test-Conditions

Order	: BRABENDER	Mixer-Temp.	: 140 °C
Operator	: EICKMEIER	Speed	: 30 1/min
Check-Date	: 16. MAY '88	Meas. Range	: 50 Nm
PL-Type	: 2000-3	Zero-Suppr.	: 0 %
Mixer-Type	: MB 30	Damping	: 3
Load. Chute	: MANUAL + 5 KG	Test-Time	: 5.0 min
Sample	: THERMOSETTING	Sample-Weight	: 24.00 g
Additive	:	Codenumber	: 1
		Start-Temp.	: 137 °C

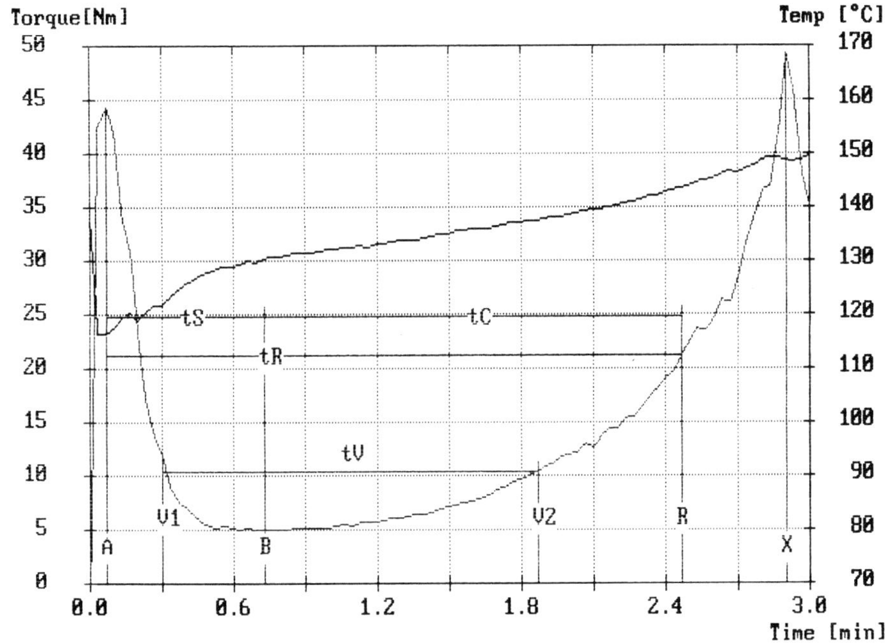

Value		at Time	Torque [Nm]	Stocktemp.[°C]
Loading Peak	A	00:00:04	44.3	117
Minimum	B	00:00:44	5.0	130
Maximum	X	00:02:54	49.3	149
Start Delay Time	V1	00:00:18	11.7	122
Stop Delay Time	V2	00:01:52	10.4	138
Stop Reaction	R	00:02:28	21.2	144

Integration / Energy

- Load.Peak to Minimum	A - B :	W1 =	1.6 [kNm]
- Minimum to Reaction	B - R :	W2 =	3.1 [kNm]
- Delay Time	tV :	W3 =	1.9 [kNm]
- Load.Peak to Reaction	A - R :	W4 =	4.7 [kNm]
- Specific Energy (W4/Sample-Weight)	:	W5 =	0.2 [kNm/g]

Results

- Delay Time	f(B+ 5.0 Nm)	V1- V2:	tV = 00:01:34
- Melt Time		A - B :	tS = 00:00:40
- Curing Time		B - R :	tC = 00:01:44
- Reaction Time	f(B+ 15.0 Nm)	A - R :	tR = 00:02:24

C

pound is fed to the mixing chamber, which is heated externally to either 248°F (120°C), 302°F (150°C), or 329°F (165°C).

As the measuring head rotors meet with immediate resistance torque they will cause the dynamometer to rotate in the opposite direction. This reaction torque acts on the indicator scale through a lever system of analytical precision, simultaneously recording on a strip chart. A typical thermoset curing curve as measured by the Brabender instrument is illustrated in Figure 17.

Generally the Brabender test is conducted at two temperature settings—248°F (120°C), simulating the temperature conditions within the barrel, and 302°F (150°C), simulating the actual molding condition. There are times when it may be necessary to run at 329°F (165°C) if the molding temperatures are to be higher to achieve faster curing cycles.

In most cases thermosetting compounds designed for use in screw transfer or screw injection equipment will show a Brabender time of 30–45 min at the lower (248°F/120°C) temperature while showing a sharply reduced time of 2–3 minutes at the higher (302°F/150°C) temperature. Such a flow profile would indicate that a specific "shot" entering the heated barrel at the hopper opening would have ample "life" or flow with the barrel temperature set at the 245–250°F range.

The reduced compound life as it increases in temperature due to the heat of the mold and the shear effect of the compound passing through the runner and gate will also reduce its viscosity and decrease its curing rate. This is reflected by the shortened life of the compound when the Brabender test is run at 302°F (120°C). However, the total injection time for most thermosets will be less than 10 sec, providing sufficient time for the molder to take full advantage of the more rapid cure rate available with higher mold temperatures.

The Brabender instrument can be used as a quality control device for production and for final flow determinations; it is also a valuable tool for the development of new compounds.

2.6 CURE CHARACTERISTICS

The chemical reaction that occurs when thermosetting compounds are subjected to heat, time, and pressure during the molding process will result in the compound becoming "cured," and the degree of cure can affect the ultimate mechanical, electrical, chemical, and thermal properties of the molded article. There are tests that can be made on molded parts to determine this degree of cure, and they vary according to the specific compound utilized.

There is a general visual check employed by most molders to determine the degree of cure of a molded part that is based upon the following criteria:

- Lack of sufficient rigidity when ejected from the mold
- Surface appearance:

Cracks
Porosity
Blisters/swelling
Dullness

- Dimensional problems:

Warpage
Flatness
Concentricity

Should it become necessary to conduct more sophisticated cure measurements, the following cure tests can be employed:

2.6.1 Allyl Compounds

Sections of a molded specimen can be refluxed (Figure 18) in boiling chloroform for 3 hr, after which the specimen is visually examined for evidence of serious attack, which shows up as surface cracking, fissuring, or more serious degradation.

Surface hardness (Barcol hardness test) can be utilized but must not be made until hours *after* removal from the boiling chloroform. The Barcol test should be made on the specimen before and after the chloroform test in order to properly evaluate the degree of undercure.

2.6.2 Phenolics and Epoxies

The ASTM D-494 procedure describes the acetone extraction test commonly employed to determine degree of cure of parts molded with these compounds. Again, the appearance of serious surface degradation will indicate undercure.

2.6.3 Thermoset Polyesters

The use of the Barcol hardness test, prior to and after boiling a section of the molding in water, will detect undercure.

2.6.4 Silicones and Aminos

ASTM does not include cure characteristic tests for the amino or the silicone compounds.

The above chemical tests do not accurately illustrate the effects of "undercure" on the overall mechanical, electrical, chemical, or thermal properties. Such parts, meeting the visual requirements as to freedom from blisters, war-

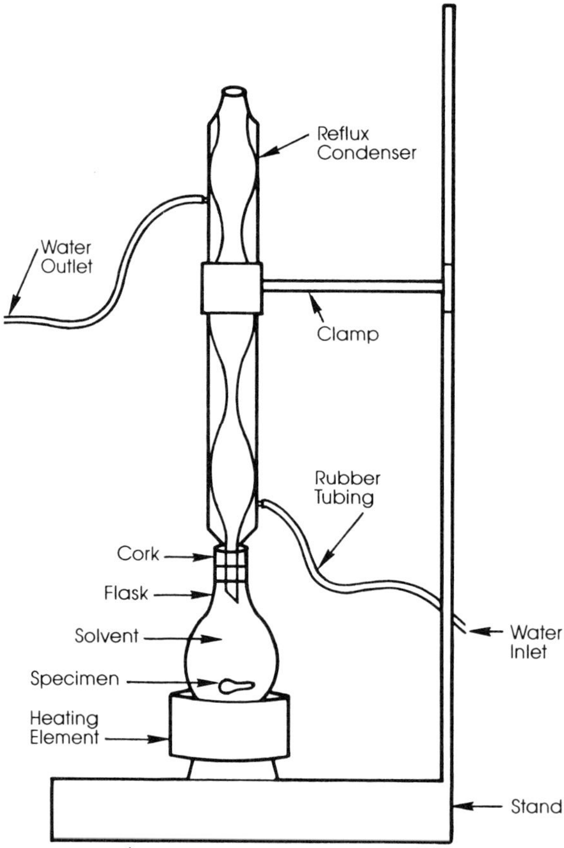

Equipment: Wide Neck Extraction Flask
Allihn Condenser
Adjustable Hot Plate
Cork Stopper
(Rubber Hose for Water Connection)

Material: Chloroform, T.G. — For Diallyl Phthalate Parts or Acetone T.G. For Epoxy Parts

Figure 18. Reflux apparatus.

page, rigidity, and dimensional correctness, will exhibit properties readily acceptable for commercial applications.

Should it become necessary to further examine the integrity of a molded part, there are more sophisticated thermoanalytical techniques available, such as

- Differential scanning calorimetry (DSC)
- Thermomechanical analysis (TMA)
- Thermogravimetric analysis (TGA)

Such equipment will analyze changes in the physical and chemical properties of a molded article as a function of temperature. These testing methods are generally used in troubleshooting situations when the standard cure tests prove inadequate.

REFERENCES

1. American Cyanamid Corp., Wayne, NJ.
2. Werner Pfleiderer Corp., Ramsay, NJ.
3. Buss (America) Inc., 2411 United Lane, Elk Grove Village, IL.
4. Premix Inc., Molding Compound Div., Kingsville, OH.
5. Hull Corp., Hatboro, PA.
6. C. W. Brabender Instruments Inc., S. Hackensack, NJ.

CHAPTER 3

MOLDING

3.1 PROCESSES AND EQUIPMENT

The molding of thermoset compounds is accomplished with three fundamental processes:

- Compression molding
- Transfer molding
- Injection molding

There are variations within each category as well as a variety of mold designs utilized in all three processes, but there does exist a commonality of purpose regardless of the process chosen. This purpose is to produce moldings that meet appropriate specifications and to do so at an economical rate of production.

The eventual choice of process will, to a large extent, be determined by the ability of the chosen process to accomplish that purpose and it becomes imperative that in the manufacturing procedure each phase of the entire molding procedure is carefully examined and documented before any mold construction is commenced. The manufacturing procedure has to include the following:

- Product design
- Compound design (emphasis on reinforcement selection)
- Molding process
- Mold Design
- Molding press selection
- Postmolding tooling and fixtures

It is not often that a single individual will possess a full understanding of each of the above facets and thus it becomes necessary to solicit needed expertise from people knowledgeable and experienced in each of those areas. The product designer, molder, mold designer, and compound supplier each has invaluable and specific knowledge to contribute to the overall process, and they should be working in concert from the beginning of the procedure. Changes or revisions to the product design, necessitating revisions to the mold, should be done in the design stage since these kinds of changes are extremely costly and time consuming when done *after* the mold construction has been completed.

3.2 MOLDING COMPOUND PREPARATION

Before discussing the actual molding processes it should be understood that the compound itself will need some preparation prior to molding—especially when using the compression or transfer methods.

In any thermoset molding operation there are several essential factors critical to the success of the operation:

- Shot weight (amount of compound needed for each cycle)
- Feeding the compound into the mold
- Temperature of the compound entering the mold
- Molding parameters

The above critical factors will depend on the molding method selected, with screw transfer and screw injection molding techniques differing from the compression and standard transfer molding methods by the fact that screw transfer and injection methods employ a "built-in" process for controlling shot weight, temperature, and compound delivery, whereas these segments of the molding cycle for compression and transfer molding are accomplished with auxiliary equipment.

In automatic compression molding the shot weight is determined by the use of a volumetric feed mechanism that feeds a carefully measured amount of either room temperature or preheated compound to the mold cavity, whereas in semiautomatic compression and transfer molding, the mold is fed with preheated preforms produced with a "pelleting" or preforming machine capable of compressing a measured amount of compound into a desired shape and weight. The preform is then heated in a high-frequency dielectric oven capable of raising the temperature of the preform from room temperature to 175–200°F (79–93°C) in a matter of seconds, ready then for entry into the mold.

3.2.1 Preforming

Preforming is a method of "cold pressing" granular molding compound into a "pill" or preform of a specific shape and precise weight. The preform provides an easy way to handle the mold charge both for preheating and for charging the mold. The preforms can be round, square, or rectangular, with the square and rectangular shapes preferred since they can be placed in the preheater on edge. This means that the preform height will be uniform from shot to shot whether loaded singly or with several preforms—a necessity in RF heating devices. Round preforms can be placed onto a pair of rotating shafts within the

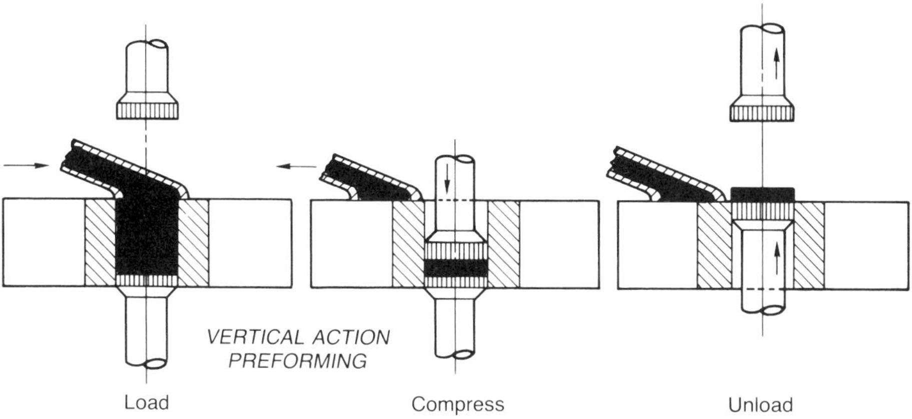

Figure 1. Rotary preformer.

preheater and the preforms will rotate on their edges during the heating cycle – accomplishing the necessary uniform height.

Modern preformers are designed as vertical or horizontal hydraulic presses that receive the compound from a volumetric feeding device and compress it into a desired shape and weight, then eject the preform for the next charge. These machines operate very rapidly at rates of 10 or more cycles per minute.

Rotary preformers are available and are usually used for small preforms where very large quantities are required. They consist of a rotating table with a mechanical cam action that provides the force necessary to compress the compound and then discharge the preform from the cavity (Figure 1).

3.2.2 Preheating

As mentioned above, the preform can be preheated to a desired temperature with the use of a high frequency dielectric heating device (Figure 2). The heating is accomplished by taking advantage of the loss factor of the molding compound. The material is heated by using a rapidly oscillating high voltage field to make polar molecules oscillate. This causes increased kinetic molecular activity, which is "heat," and the energy absorbed raises the temperature.

All portions of the material heat at the same time, but the outer surfaces of the preform lose heat to the surroundings whereas the center does not. This is why the material seemingly heats from the inside out.

RF heaters operate at 70 to 100 MHz (frequency) and RF voltages of 5 and 35 kV. The rate of heating is directly proportional to the frequency and varies as the square of the voltage imposed across the material. This is why electrode spacing in the heater (adjustable for different preform heights) profoundly affects the heating rate. Preform rotators, used with round preforms, improve

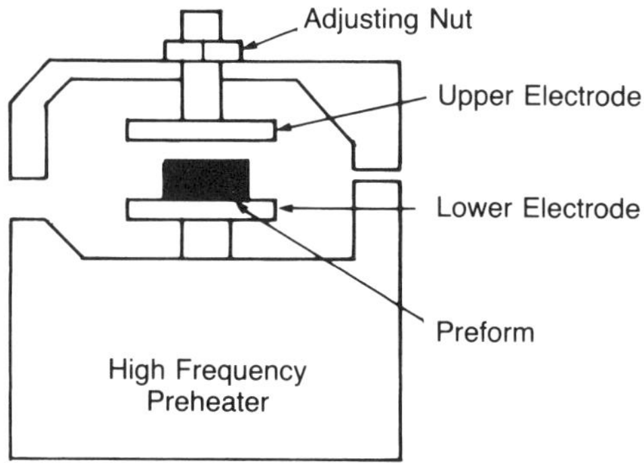

Figure 2. High frequency dielectric heating device.

the uniformity of heating in a preform at a small sacrifice in heating rate. The rate of heating increases as the temperature increases, so a change of a second or two in heating time will significantly affect the temperature attained.

Some causes of uneven RF heating include

- Uneven preform height
- Electrodes that are not parallel
- Sharp edges on the preforms
- Varying preform density
- Slight differences in preform moisture content

Note: The author is grateful to David Daniels of the Rogers Corp. for his contributions to the dielectric preheating process.

3.2.3 Preplasticating

Another piece of auxiliary equipment for providing a measured amount of heated compound is the preplasticator, of which there are two versions.

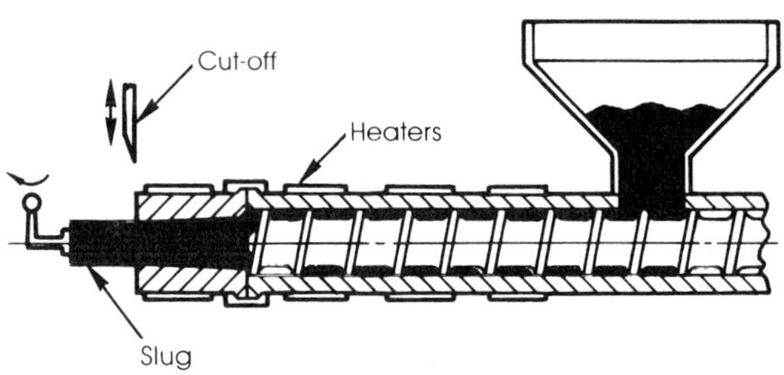

Figure 3. Nonreciprocating screw plasticator.

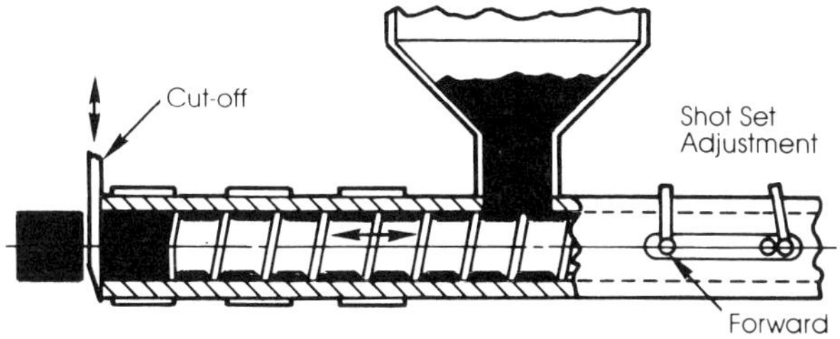

Figure 4. Reciprocating screw plasticator.

As can be seen, nonreciprocating the machine (Figure 3) is an extruder with a nozzle at the exit end that contains a tapered opening through which the heated compound is forced and compressed by a stationary, rotating screw into a preset size. The extrudate is forced out of the nozzle until it comes in contact with a limit switch, which, when depressed, activates the cut-off knife, producing a heated "slug" of compound of a desired weight and temperature. Both the weight and temperature are very consistent.

The reciprocating machine (Figure 4) has no nozzle and the inside diameter of the barrel determines the diameter of the extrudate. With the cut-off knife in the "down" position, the screw rotates and moves backward, against a preset back-pressure, forcing the heated compound against the face of the knife, densifying the compound. As the screw reaches the preset "shot adjustment," the limit switch causes the screw to stop and the cut-off knife to return to its "up" position. On reaching its up position, another limit switch is activated, causing the screw to move forward under pressure, acting as a plunger, and forcing the extrudate from the barrel. The screw, moving toward the exit end of the barrel, will contact the forward position of the shot adjustment, activating the cut-off knife, which will then cut off the desired weight slug, ready for delivery into the mold.

Either preplasticator, reciprocating or nonreciprocating, is capable of being programmed into the molding cycle and will produce "slugs" of extremely uniform weight, density, and temperature.

The advantages of the preplasticator over the preform/preheater system are

- Combining of the preforming and preheating operations
- Elimination of the RF heaters that are expensive and costly to maintain
- Producing extrudates of more uniform temperature and density, resulting in less scrap and faster curing cycles

Preplasticators or thermoset extruders are also utilized by the suppliers of thermoset polyester compounds when the supplier ships the compound in a "rope" or "log" form, which the molder can size for the specific molding operation. The sized shot charge can be preheated for semiautomatic transfer or compression molding or can be fed to an automatic transfer molding setup as depicted in Figure 5.[1]

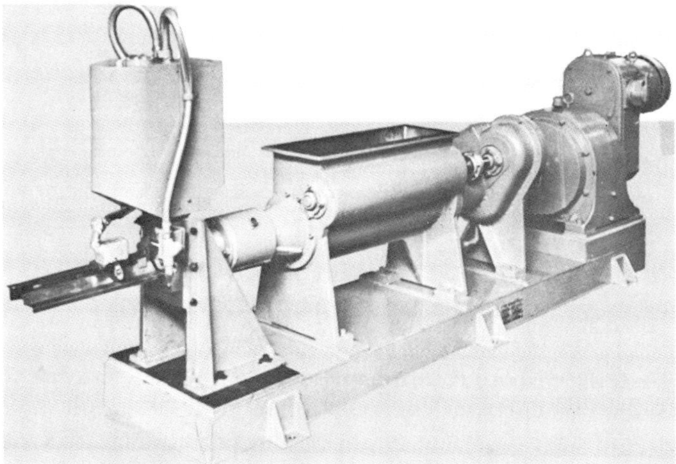

Figure 5. Thermoset extruder preformer.

There are instances when compression molds are manually fed with either room temperature compound or loose compound, which is preheated with infrared lamps or heating ovens beside the molding press. This is usually done when loading the cavity with loose compound to achieve maximum molded density or when it becomes necessary to "breathe" the mold. The mold is closed and then reopened and reclosed. This breathing action helps dissipate any volatiles and is especially effective when molding very intricate and/or very deep parts.

With screw transfer and screw injection equipment the entire preheating and weight control process is accomplished within the molding machine itself. This will be covered in more detail in Chapter 4.

3.3 MOLDING PROCESSES

3.3.1 Compression Molding

Compression molding employs either a hydraulically operated press (Figure 6) or a press utilizing hydraulically actuated toggles.

The press consists of two mold platens, one that is stationary and one that moves, either upward or downward, depending on the press design. The platens have either "T" slots or tapped holes that serve as a means for clamping the mold halves to the platens. The tie rods—generally four in number—provide for precise alignment as the press is opened and closed. The press open and shut height is controlled by the adjustment of the large nuts located under the lower platen and over the upper platen. In hydraulically actuated straight ram presses, the press closes until the mold faces come into contact.

Compression presses are rated by their closing force capacities referred to in tonnage and range from the smaller 6 ton to very large machines capable of

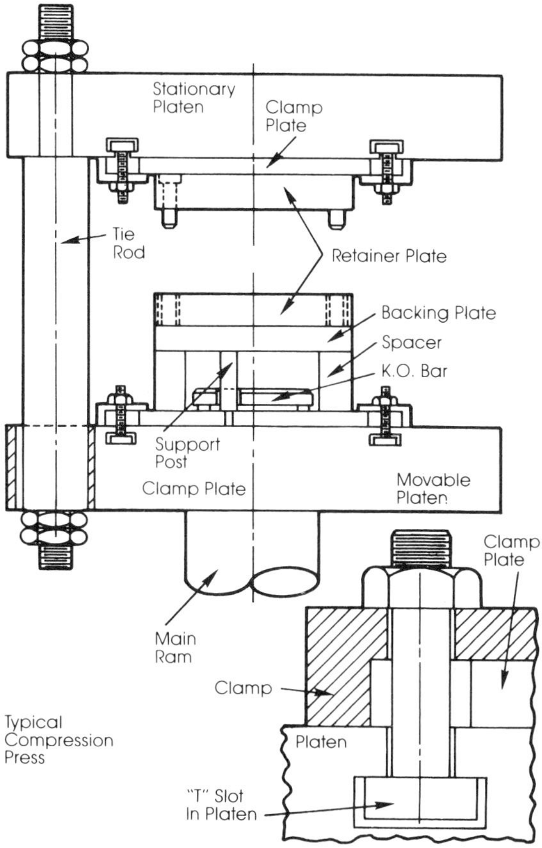

Figure 6. Compression molding with a hydraulically operated press.

providing as much as 5,000 tons or more of pressure. Most of the compression presses operate semiautomatically, but many do have fully automatic capabilities.

For a fairly long time, prior to and after World War II, fully automatic compression molding presses—mostly in the 15–75 ton range—became very popular for the production of high-volume, low-cost articles. These machines are programmed to cycle cold granular compound from a hopper attached to the rear upper portion of the press. A "loading" plate containing holes matching the cavity locations in the mold is positioned under the hopper, where the holes are filled with the proper weight of compound, and then the plate slides forward, over the cavities, delivering the compound to each cavity and then retracting to its position under the hopper for refilling. The press then closes, the parts are cured, and the mold opens with the molded parts being blown through a "trap door" at the rear of the press into a collection bin, completing the cycle. The trap door serves as a safety device controlling the beginning of the next cycle. If the parts do not go through the trap door, the press will not commence

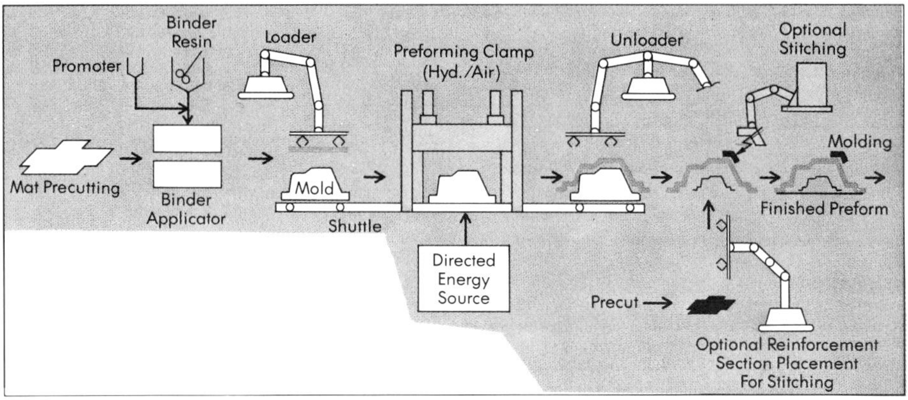

Figure 7. Compform directed-energy preforming press.

another cycle, forestalling any mold damage from "double shots." A signal sounds when this occurs and the press stops until the operator corrects the situation and clears the mold for the next cycle. Other automatic compression presses make use of "stripper combs" to remove the molded parts from the ejector pins and deliver the molded parts to a bin located at the rear of the press.

This was an economical procedure with a single operator attending several presses but, with the use of cold compound, the molding cycles were in the 3–4 min range. Using preheated compound was of some aid, but was difficult to control.

With the advent of the newer thermoset polyester molding compounds (SMC, BMC, and TMC), which are usually compression molded, it has become necessary to utilize press designs that are more suitable for molding these newer compounds. Because these compounds tend to cure rapidly and, at the same time, must remain in position as the mold closes, the press closing sequence has to be very carefully controlled and the press itself must not allow any flexing of the platens at closing or during the opening and molded part ejection. Such demands on the press brought about innovations both to the molding press and also to the technique of loading the compound into the mold to ensure proper placement of the compound "blanks" or sheets prior to mold closing. The presses were built with sturdier tie rods and thicker platens to achieve the needed stability, and the closing and opening stroke distances were shortened to reduce the time needed to close. Process controls were initiated, allowing the programming of the entire close to open cycles along with an automated blank loading system. In all, these process improvements led to an ever-increasing utilization of these compounds in many very large area moldings (Figure 7).[2]

When using the compression molding process it must be understood that the cavity depth is an important factor in determining the amount of molding pres-

Table 1. Pressure, psi of Projected Land Area

Depth of molding (in)	Conventional phenolic		Low-pressure phenolic	
	Preheated by high frequency	Not preheated	Preheated by high frequency	Not preheated
0–¼	1000–2000	3000	350	1000
¼–1½	1250–2500	3700	450	1250
2	1500–3000	4400	550	1500
3	1750–3500	5100	650	1750
4	2000–4000	5800	750	2000
5	2250–4500	*a*	850	*b*
6	2500–5000	*a*	950	*b*
7	2750–5500	*a*	1050	*b*
8	3000–6000	*a*	1150	*b*
9	3250–6500	*a*	1250	*b*
10	3500–7000	*a*	1350	*b*
12	4000–8000	*a*	1450	*b*
14	4500–9000	*a*	1550	*b*
16	5000–10000	*a*	1650	*b*

[a]Add 700 psi for each additional inch of depth, but beyond 4 in. in depth it is desirable (and beyond 12 in. essential) to preheat.
[b]Add 250 psi for each additional inch of depth, but beyond 4 in. depth it is desirable (and beyond 12 in. essential) to preheat.

sure required to achieve the proper molded density. Table 1[3] provides some examples of required molding pressures when using both conventional general purpose phenolic compounds and low-pressure, soft-flow phenolic compounds. The pressures shown will also apply to most thermosetting compounds of the granular type whereas the bulkier, high-strength compounds will require higher molding pressures with further increased pressure as the cavity depth increases.

3.3.2 Transfer Molding

3.3.2.1 Pot Type Transfer Mold

As can be seen, the mold design (Figure 8) provides a "pot" into which the compound, heated or unheated, is placed and the press closed. The clamping pressure exerted by the press also serves to force the compound from the pot into a runner system and then into the cavity. The bottom of the plunger is undercut to serve as means for pulling the cull (waste material) from the pot as the press opens at the mold parting line for ejection of the molded parts.

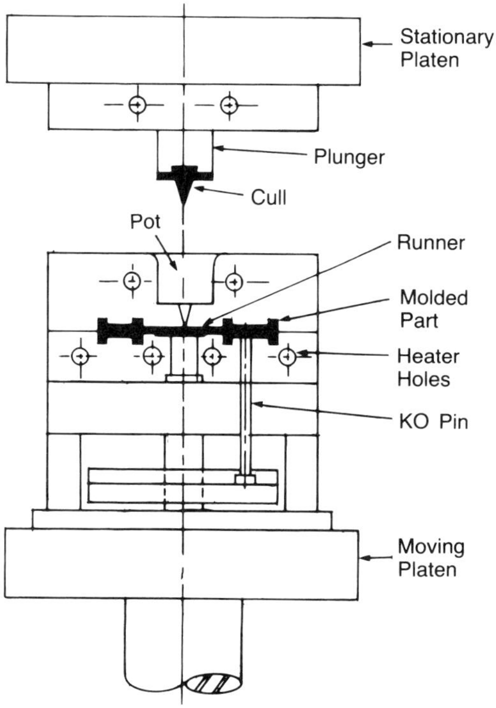

Figure 8. Pot-type transfer mold.

Pot transfer molding was an innovation providing an improved means for molding parts with metal inserts as well as parts with less flash at the parting line and in through-molded holes. It also made it possible to feed the molding compound into a single location, rather than into each separate cavity as is often the case with semiautomatic compression molding. However, the entire operation proved to be quite clumsy and resulted in an extremely large quantity of waste material, since it was necessary for the area of the pot, itself, to be larger by some 15% than the combined area of the cavities and the runner system to overcome the pressure of the compound as it entered the runner system and the cavities, thus keeping the mold completely closed. The resultant high cost of waste compound forced the molders to seek a more economical method of molding with a closed mold.

3.3.2.2 Plunger Transfer Molding

An innovation to the molding procedure came into being in the late 1930s and early 1940s to overcome the difficulties of the pot type molding method. The

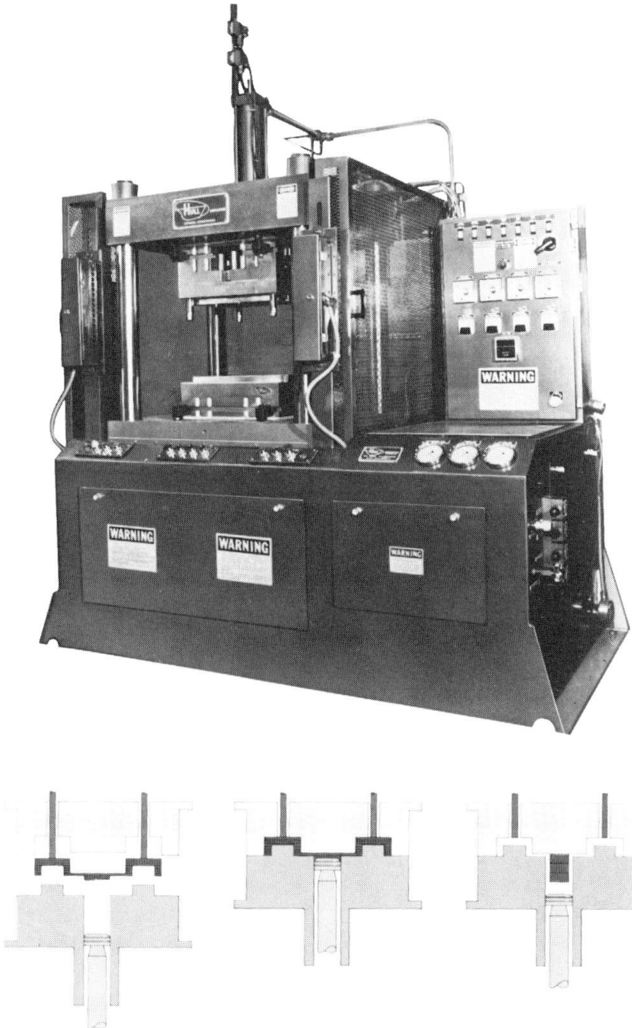

A

B

Figure 9. (A) Top ram transfer molding press. (B) Transfer molding press schematic.

rather simple act of mounting a hydraulic cylinder onto the top platen of the standard compression press gave birth to what has been called "plunger" or high-speed transfer molding. As can be seen in Figure 9[4] clamping pressure exerted by the lower platen is utilized to keep the mold totally closed while the top cylinder provides the pressure required to force the compound into the mold. This technique, along with the advent of the high frequency dielectric preheater, enabled the molder to take advantage of the ability to feed uniformly heated preforms to the mold and have the mold produce parts with little or no

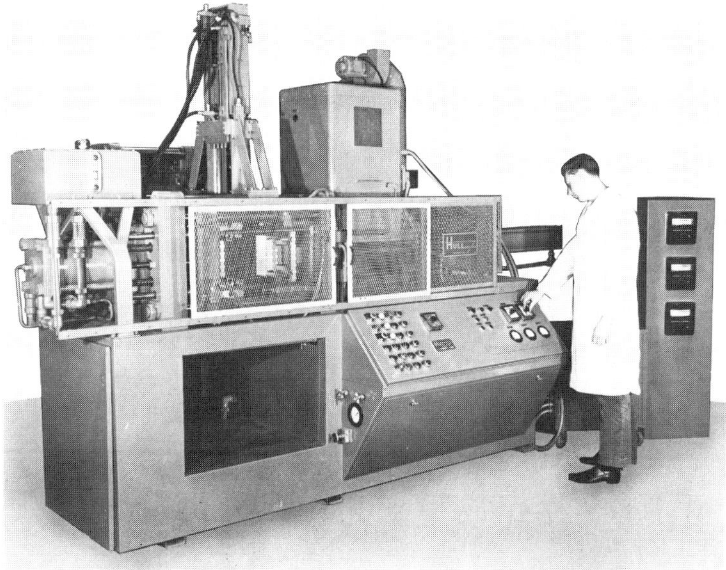

Figure 10. Fully automated transfer molding machine with integral preforming and high frequency preheating (courtesy Hull Corp.).

flash (waste) on the parting line or in any molded-in holes. It also replaced the old pot with its large cull and sprue waste.

Some of the other advantages to plunger molding are

- Greater hole depth/length ratio
- Faster cycle times with higher productivity
- Increased ability to mold undercuts with side draws
- Improved molded density
- Minimum compound waste
- Less flash
- Reduced finishing costs
- Close control of dimensions perpendicular to the mold parting line

3.3.2.3 Automatic Transfer Molding

In an attempt to incorporate preform/preheating as an integral part of the press function, the Hull Corporation provided a designed and built molding press containing an overhead preformer capable of feeding a precise weight preform to a dielectric preheater directly beneath it. The preheater would then preheat the preform and feed it to the plunger "barrel" and then into the mold via the plunger ram (Figure 10).[4]

Figure 11. Model TS 350/1810 with CNC controls.

This molding technique was capable of completely automatic cycling and enjoyed considerable success from the mid 1950s thru the 1960s.

The automatic transfer molding as described above was confined to the use of free-flowing granular compounds but, with the advent of the high-strength, high-bulk factor thermoset polyesters, it became necessary to redesign the molding press to accommodate these newer compounds and to take advantage of the lower cost, faster cycles of automatic transfer molding.

The initial press designs utilized a feed barrel or reservoir that was fed with precut, preweighed logs from a rotating carousel or "gatling gun" arrangement.

The plunger/transfer molding machines (Figure 11)[5] are currently used extensively for the molding of electrical switchgear, wiring devices, and appliance components with polyester BMC and TMC compounds because the plunger action is less destructive to the reinforcing fibers, producing moldings with superior physical properties.

The inherent problems encountered with this design revolved around the following considerations:

- Inability to evacuate volatiles from the stuffing unit
- Need for a seal between the plunger and barrel to eliminate back flow during the transfer cycle
- A "cushion" of compound had to be retained in the barrel and kept at a uniform temperature to insure consistency in the molding operation

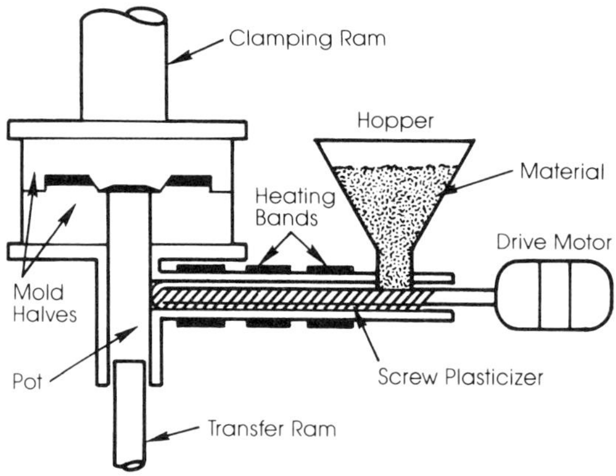

Figure 12. (A) Stokes screw transfer press (courtesy Stokes Company).

These problems led to the use of a screw injection process that will be described below.

3.3.2.4 Screw Transfer Molding

In the ongoing evolution of thermosetting molding a novel concept came into being through the Stokes Co., a long time manufacturer of automatic compression molding machines. This concept served to eliminate both the preforming and the preheating functions as separate operations outside of the press itself (Figure 12).[6]

This new molding technique with its use of an extruder could supply a precise weight, temperature, and density "shot" and then feed it into the mold on a completely automatic basis.

Again, the Stokes Injectoset molding press was confined to the use of free flowing granular compounds and was not intended for the bulky, high-strength compounds.

3.3.3 Injection Molding

Automatic and screw transfer molding equipment advances in thermoset molding techniques were the forerunners of the next innovation, which took advantage of both ideas, resulting in the design and production of what became known as the in-line screw press.

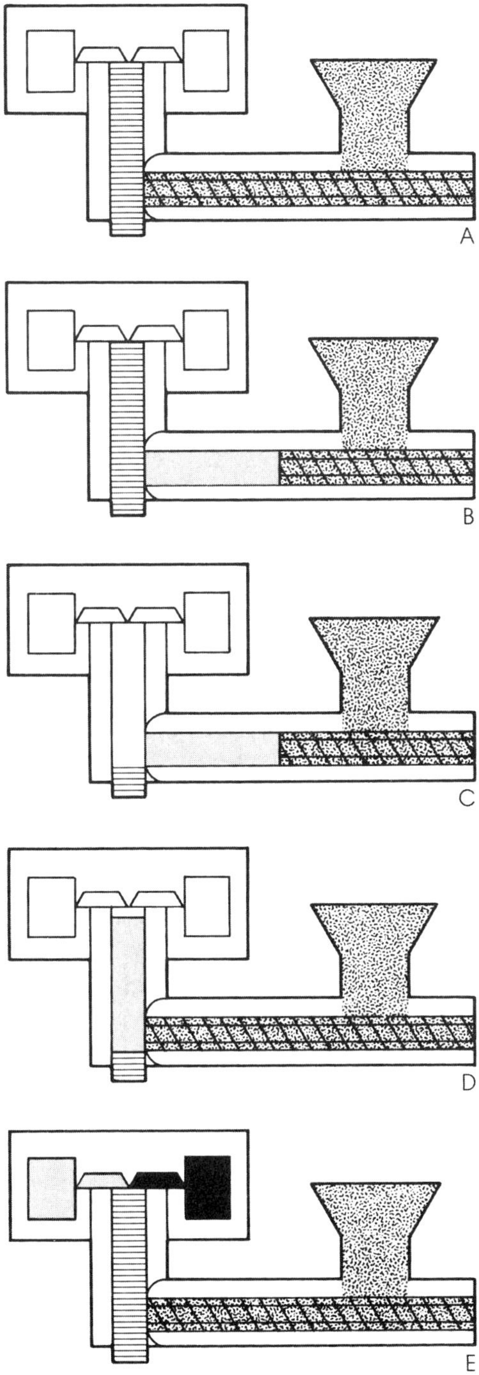

Figure 12. (B) Stokes screw transfer process (courtesy Stokes Company).

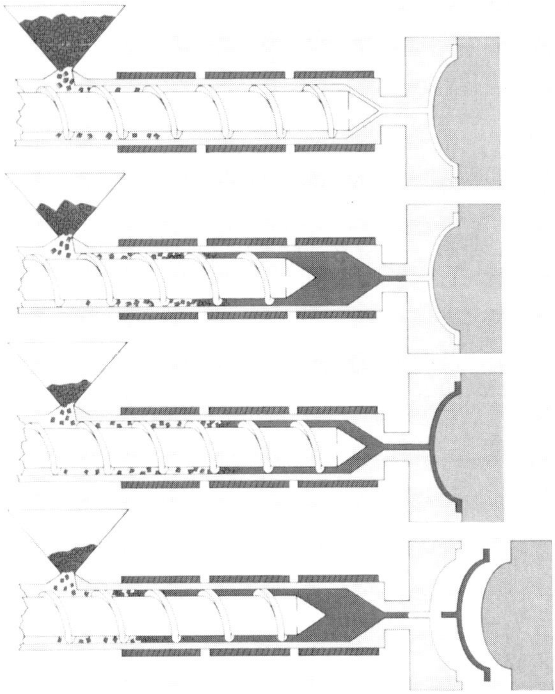

A

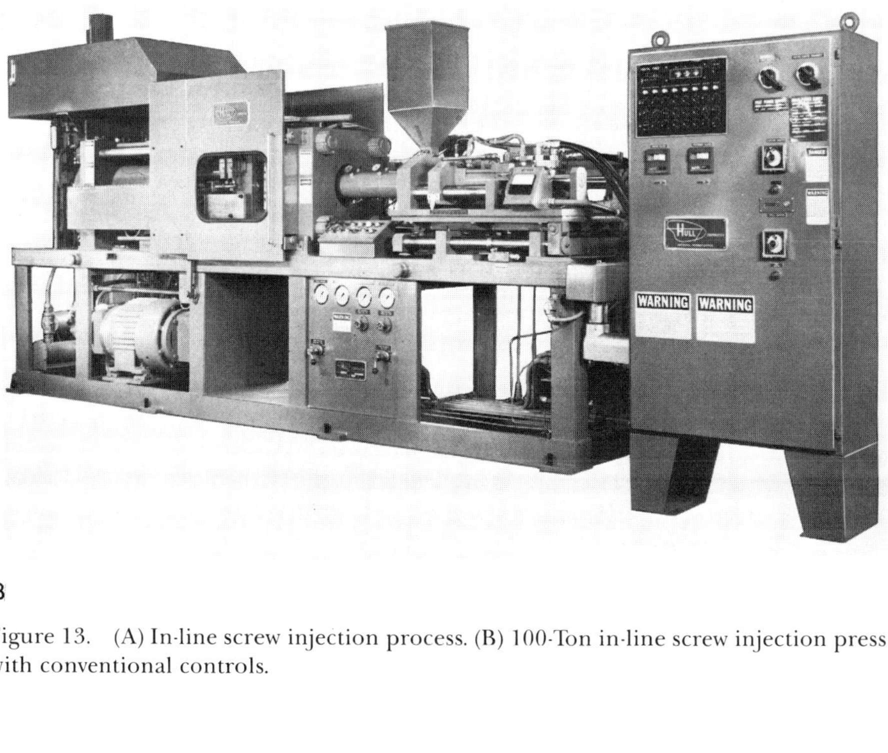

B

Figure 13. (A) In-line screw injection process. (B) 100-Ton in-line screw injection press with conventional controls.

C

Figure 13. (C) In-line screw injection press with computer controls.

3.3.3.1 In-Line Screw Injection Molding

As can be seen in Figure 13[4] the extruder unit becomes an integral part of the molding press with the extruder providing the preheated mold charge and the screw itself becoming the plunger unit—all on a horizontal plane with the molded shot being ejected from the mold and falling beneath the the mold onto a conveyor mechanism or into a bin. The fundamental differences between the screw transfer press and the in-line screw press are the plane on which they operate and the speed with which the heated mold charge is fed into the mold. With the screw transfer press there is a time delay as the plunger retracts itself to the down position to receive the heated mold charge and then proceeds to force the compound into the closed mold. With the in-line press the next shot is processed in the extruder just prior to the mold opening when the parts are ejected. Immediately on closure of the mold the screw/plunger is activated, forcing the next heated charge into the mold.

There is a time savings of approximately 15 sec with the in-line press, resulting in shorter production cycles and lower costs.

The system performs very well with free flowing granular compounds but requires several modifications to handle the bulky, high-strength thermoset polyester compounds.

For such materials, there is need for a revised screw design with deeper flights, as well as a reverse flow check valve to prevent any back flow of compound during the high-pressure injection cycle. During the curing phase of the molding cycle, the screw reciprocates, preparing the next shot. As this occurs, the stuffer is force feeding the compound from the stuffer onto the flights of the rotating screw. Thermoset polyesters are rapid curing compounds and their

Figure 14. Plunger/injection with stuffer systems for automatic feed (courtesy Hull Corp.).

Figure 15. Parting/line injection press (courtesy Hull Corp.).

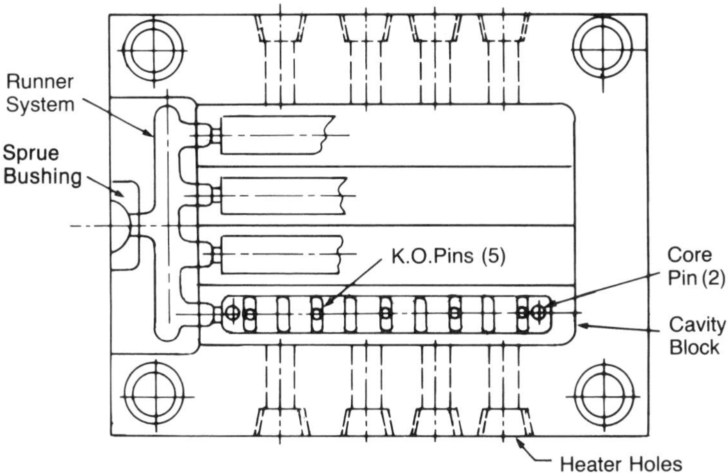

Figure 16. Plan view of a 4-cavity parting/line mold.

barrel residence time is relatively limited compared with other thermosets (Figure 14).[4]

In deciding the merits of the plunger/transfer (PT) vs. the screw injection (SI) the following concerns need investigation:

- Fiber degradation—generally more severe with SI.
- SI requires little or no cushion, making shot control more accurate, resulting in less flash and more consistent moldings.

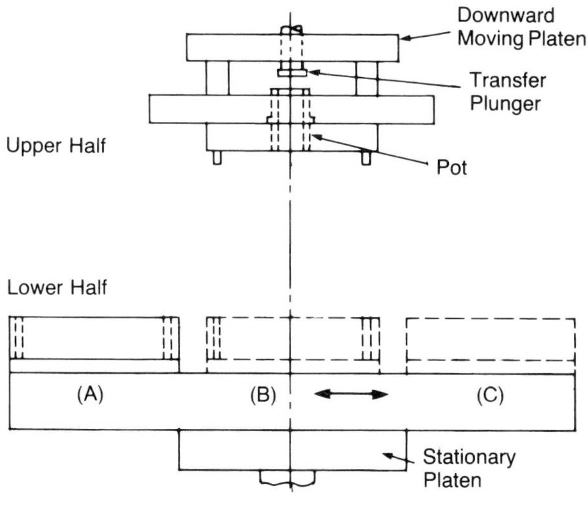

Figure 17. Shuttle press for insert molding.

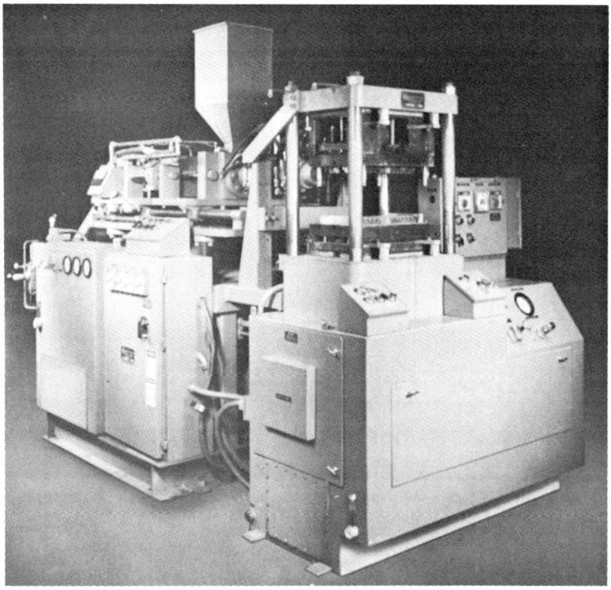

A

B

Figure 18. (A) Parting line injection module coupled to a Hull 359-F, fifty-ton press. (B) Shuttle press (courtesy Hull Corp.).

Figure 19. Special 150-ton shuttle-type transfer molding machine.

- SI molding cycles are faster due to the ability to preheat the compound prior to injection.
- Surface finish is of higher quality with SI.
- Compounds with lower glass loading are better suited for the SI process, whereas the higher glass loaded compounds will prove stronger when molded in the PT process.

3.3.3.2 Parting Line Injection Molding

With product designers taking more and more advantage of the ability to include molded-in metals inserts in their part designs, it became apparent that the horizontal machine presented certain problems relating to both the loading and the retention of these inserts during the molding process. The vertical press with its horizontal platens provided a solution to those problems, and the molding machine designers accommodated the molders with a screw injection unit that came in from the side of the press and could be adjusted vertically so that the nozzle of the unit was located at the parting line of the mold (Figure 15).[4]

Another advantage of this type of press is when molding long parts as shown in Figure 16.

3.4 SHUTTLE PRESS MOLDING

In the use of molded-in inserts, vertical presses with horizontal platens are predominantly employed because of the ease of loading the inserts into the mold along with the safety factor of being able to retain the inserts in their proper location during the opening and closing of the mold.

3.4.1 Transfer Shuttle Press Molding

A popular transfer press design for insert molding is the shuttle press (Figures 17, 18,[4] and 19[4]) in which the top half of the mold is mounted to the upper press platen and contains the pot. The lower press platen will have two lower mold halves mounted to it. This platen will move laterally via hydraulic rams to properly position the lower mold halves when they are molding or are in the unload/load position. The unload/load stations are also equipped with hydraulic systems to actuate the ejector plates in each lower mold half.

3.4.2 Injection Shuttle Press Molding

When utilizing the injection process with a shuttle concept, the parting line press, with its vertical clamping and horizontal injection units, lends itself very favorably to the process. The lateral movement of the lower press platen provides for the unload/load operations.

A similar process, where the mold, with two lower halves, can be used with the operator removing lower half #1 from the press platen onto a heated plate outside the press itself, where it is unloaded and reloaded while mold half #2 is in place on the press platen and molding. This concept does not require a movable platen and can be easily set up with transfer or compression molding presses.

The choice of molding procedure for any specific part design will depend upon some of the following factors:

	Type of molding process		
Factor	Compression	Transfer	Injection
Compound load	Mostly semiautomatic	Mostly semiautomatic	Automatic
Preheating	Mostly semiautomatic	Mostly semiautomatic	Automatic
Note: Compound loading and/or preheating can be automatic with either compression or transfer with proper design and sufficient volume of moldings to cover the extra cost.			
Molding temperature	325–400°F (163–205°C)	335–410°F (168–210°C)	325–415°F (163–448°C)
Note: Generally speaking, transfer and injection molding is done at higher molding temperatures due to the closed mold setup, allowing speedier compound loading and movement of the compound into the cavity.			

Factor	Type of molding process: Compression	Transfer	Injection
Molded density	Best	Good	Good
Required molding			
Pressure	2000–8000 psi	3500–6000 psi	3000–5500 psi
Cure time	30 sec–5 min	25 sec–1.5 min	15 sec–1.5 min

Note: Cure times are highly dependent on the part configuration, molding compound and the ability to mold at maximum temperatures.

Factor	Compression	Transfer	Injection
Insert molding	Very limited	Excellent	Excellent
Use of side draws	Limited	Excellent	Excellent
Molding tolerances	Good, except off P/L	Very good in all directions	Very good in all directions
Shrinkage	Least	*Note*: Shrinkage occurs in transfer and injection molding to a greater extent than with compression and will be more pronounced in the direction of flow than in the opposite direction.	
Finishing	Most costly	Least costly in either transfer or injection	

REFERENCES

1. The Bonnot Co., Kent OH.
2. C.A. Lawton Co., DePere, WI.
3. Joel Frados, *Plastics Engineering Handbook*, 4th ed., 1976.
4. Hull Corp., Hatboro, PA.
5. Bucher Inc., Arlington Heights, IL.
6. Pennwalt Corp., Plastics Dept., Philadelphia, PA.

GLOSSARY OF THERMOSET MOLDING TERMS*

A—Stage· An early stage in the reaction of certain thermosetting resins, in which the material is still soluble in certain liquids and fusible.

B—Stage· An intermediate stage in the reaction of certain thermosetting resins, in which the material swells when in contact with certain liquids and softens and flows under heat and pressure. The resin in an uncured thermosetting molding compound is usually in this state.

C–Stage- The final stage in the reaction of certain thermosetting resins, in which the material is relatively insoluble and infusible. The resin in a fully cured thermosetting molding is in this stage.

Blister- Undesirable rounded elevation of the surface of a plastic whose boundaries is indefinitely outlined, somewhat resembling in shape a blister on the human skin. A blister consists of trapped air, water or solvent and can be caused by inadequate curing time, temperature or pressure.

Bulk Density- The density of a molding material in loose form (granular, nodular, etc.) expressed as a ratio of weight to volume (e.g. g/cm^3 or lb/ft^3).

Bulk Factor- The ratio of the volume of the loose molding compound to the volume of the same quantity in a molded solid piece. Also the ratio of the density of the solid plastic object to the apparent density of the loose molding powder.

Cavity- That portion of the mold which forms the outer surface of the molded article.

Chase- The main body of the mold which contains the molding cavity or cavities, or cores, the mold pins, the guide pins, etc.

Core- That portion of the mold that forms the required inner surfaces of the molded article.

Core Pin- Pin used to mold a hole.

Cull- Material remaining in the transfer pot after the mold has been filled.

Cure- To change the properties of a plastic by chemical reaction which may be condensation, polymerization or addition; usually accomplished by the action of heat or catalyst or both, with or without pressure.

---

*Courtesy of Horst W. Rauhut of Hysol Division, Dexter Corp., Olean, N.Y.

CHAPTER 4

MOLDING TECHNIQUES

4.1 IMPORTANCE OF MOLDED DENSITY

Regardless of the type of molding compound, molding method, or type of molding equipment employed, the basic objective is to produce, at an economical rate, sound moldings that are capable of meeting all appropriate specifications—generally imposed by the product designer. The key factor in attaining this objective is to carefully monitor and control those molding parameters directly affecting the "filling" or "packing" of the mold cavity during the molding cycle so as to achieve *maximum molded density* (MMD) along with a well-cured molded part.

The molding parameters in need of both monitoring and controlling are

- Shot weight—temperature—density
- Molding pressure and temperature
- Compression closing speed
- Transfer or injection speed

The important and essential need is to have these parameters constant and uniform from cycle to cycle with a system that will warn of deviations early enough that the proper adjustments can be quickly made.

The increasing ability of today's computer systems to provide means by which the molding process can be monitored and controlled has been, and will continue to be, a major factor in the continued growth of the thermosetting molding industry.

4.1.1 Process Controls

There have been a number of process control systems developed for the transfer and injection processes over the past several years with only recent adaptation to the compression process control over the basic molding press functions that have been in place for years but were only a beginning of real process control. The fundamental need for consistent performance of the press and the ability of the press to compensate for inconsistencies encountered by the press as the polymer reaction advances have led to the use of the control equipment in locating the sources of those problem areas. This has brought about press component design improvements susceptible to computer-aided process controls.

Total process control has not, as yet, become perfected, but there have been, and will continue to be, strong advances toward that goal. The parametric analysis database in Figure 1[1] portrays a graphic picture of the areas that have to be monitored and controlled, particularly in the injection molding process.

The achievement of MMD on a consistent basis requires constant monitoring of the entire molding process and its individual parameters along with the proper mold design and construction. The mold and the process are co-partners in the production of sound moldings. Once the mold has been brought into its best working order and proven capable of producing acceptable parts, it will remain so if properly maintained. However, the process would seem to be susceptible to many factors that could result in faulty moldings. Here is where the process controller comes into the picture and it generally commences with the functions of the molding press in conjunction with the molding cycle.

The process variables presenting themselves to the molder vary according to the process and the mold design in use. The following is a breakdown, by process, of those variables to be taken into consideration:

Compression molding:

- Mold design
- Compound composition
- Shot weight
- Preheat temperature
- Mold temperature
- Mold closing time
- Cavity pressure
- Mold breathing cycle
- Cure time

Transfer molding:

- Mold design
- Compound composition
- Shot weight
- Preheat temperature
- Transfer pressure and speed
- Mold temperature
- Runner and gate configuration
- Dynamic pressure (shear)
- Cure Time

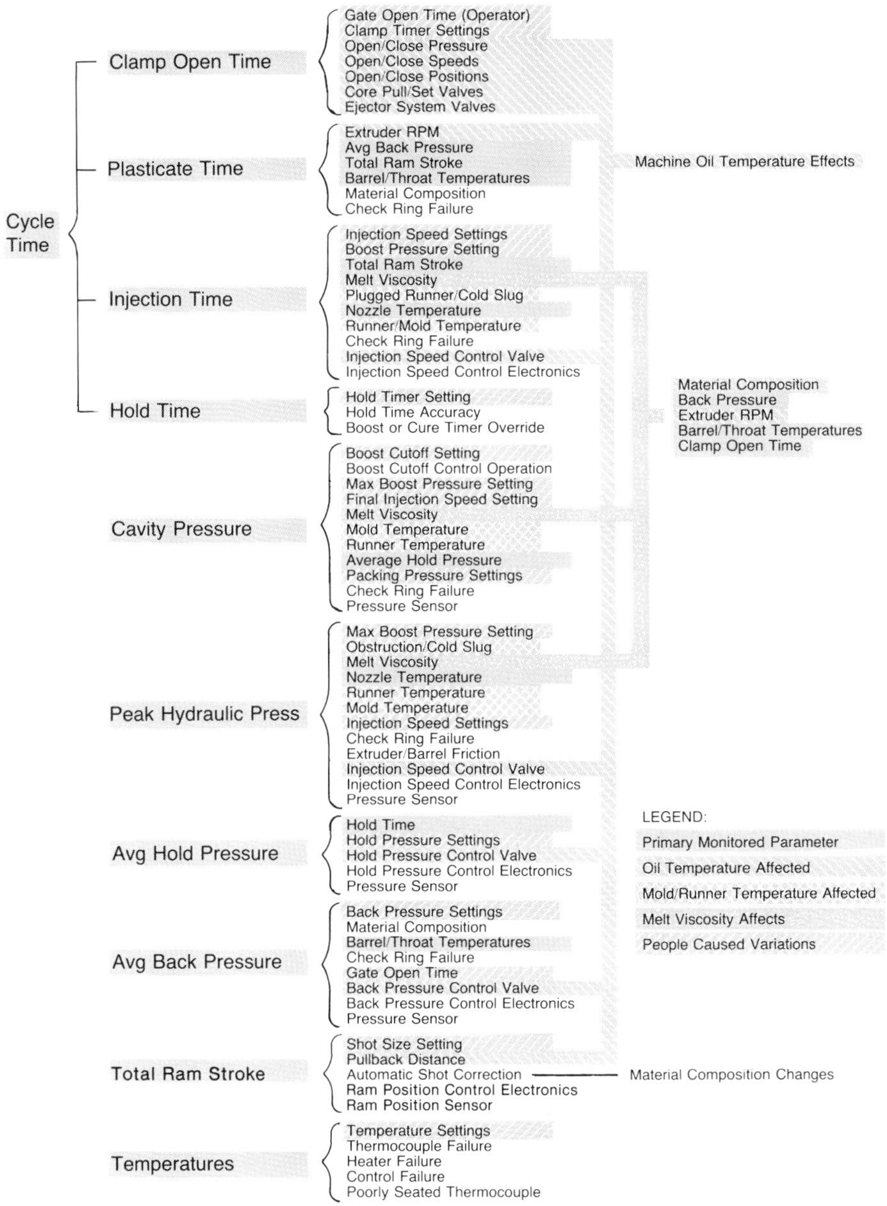

Figure 1. Parametric analysis database for statistical process control in injection molding.

Injection molding:

- Mold design
- Compound composition
- Screw rpms
- Back pressure on screw
- Dynamic pressure (shear)
- Barrel temperature
- Barrel residence time
- Nozzle temperature
- Nozzle residence time

The interrelationship of molded part problems can be traced to irregularities in the following:

- Curing or holding pressure
- Fill or "boost" pressure
- Mold temperature
- Fill or closing speed
- Shear
- Runner and gate configuration
- Shot weight consistency
- Back pressure
- Screw rpms
- Barrel and nozzle temperatures

Consistent molded part quality is highly dependent on achieving a constant curing temperature and pressure, with shear action being the primary contributor to rapid elevation of compound temperature, possibly causing precure and nonfilling. Thus it becomes incumbent that those elements directly affecting the temperature and viscosity of the compound, as it enters the mold, be brought under constant control.

It is recommended that the molding machine be equipped with a position transducer covering the injection or transfer ram along with a cavity pressure transducer installed behind one of the ejector pins in the mold. With this method of operation, the control continually adjusts the shot volume to compensate for the effects that can cause an increase or decrease in the cavity pressure during cure, taking into consideration all variables at the same time.

Other process factors to be kept constant are the screw RPMs, back pressure, and injection speed. Back pressure and injection speed are controllable with the use of a hydraulic servo valve controlling the oil pressure and flow volume in the injection unit.

With or without a process control system, it is possible to easily determine whether MMD has been achieved. A careful examination of the molded part once it has been ejected from the mold will quickly reveal defects such as listed below if MMD has not been achieved:

- Surface porosity
- Surface blisters
- Excessive part shrinkage
- Warpage affecting flatness
- Cylindrical parts out of round
- Lack of physical strength

Note: Many of these same defects are also indicative of a *lack of cure* as well as a lack of *MMD.*

Less visible, but easily determinable, evidence of a lack of MMD is an overall reduction of mechanical, electrical, chemical, and thermal properties. The achievement of MMD in any molded part will always bring out the very best property values possible from a given molding compound or molding process.

The choice of *molding process* may well revolve around the ability to achieve MMD with the any chosen compound, along with other considerations such as

- Part configuration
- Insert use
- Number of cavities and their location in the mold
- Parting line location
- K.O. pin locations
- Runner size and configuration
- Fiber orientation
- Gate location and configuration
- Mold venting

The following will examine some of the pros and cons of the three major molding methods.

4.2 COMPRESSION MOLDING

All compression molding techniques, whether flash, positive, or semipositive, rely on an overload of compound into the cavity, the use of direct pressure within the cavity, and allowance for the excess compound to escape from the cavity under controlled conditions to achieve MMD. A properly designed and constructed compression mold serves that purpose very well, especially on parts of fairly simple configuration and on parts that are deep and difficult to fill out with either transfer or injection molding. The compounds used are generally stiff in flow, requiring high pressures to fill the cavity. Cure times are in the 1–3 min range and the parts will have "flash" or excess molding

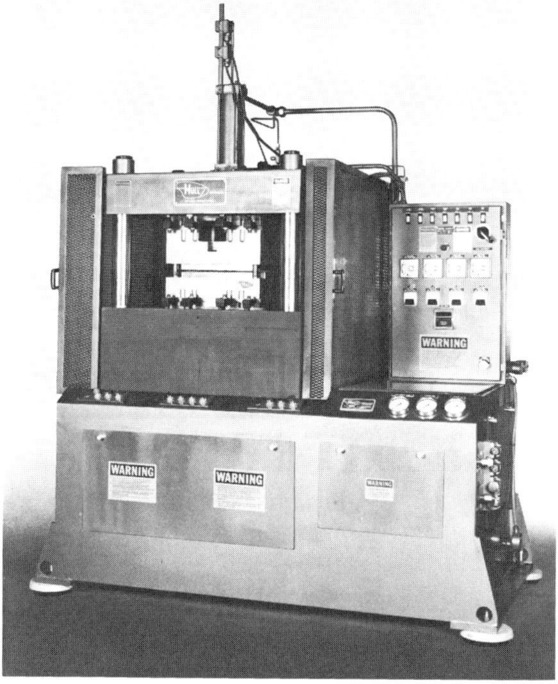

Figure 2. Transfer molding press with a top ram press cylinder.

compound attached around the parting line periphery and in any through molded holes, the removal of which will require either manual or mechanical deflashing.

It is very difficult to mold in inserts or undercuts with this mode of molding, but there are certain part designs that, due to their complex configuration and design, can *only* be compression molded, and thus the technique will continue to serve its purpose in the industry.

4.3 TRANSFER MOLDING

Transfer molding has been a favorite technique for the rapid production of thermoset molded parts since the early 1940s, when the compression molding presses were revised to include a separate transfer cylinder for forcing the compound into a closed mold. Presses were designed as shown in Figure 2[2] with a top ram transfer cylinder mounted on the top platen. An alternate press design utilizes a transfer cylinder attached to the lower, stationary platen for ease in loading the preheated preform into the mold. With bottom transfer molding, the upper platen becomes the moving platen.

Either type press provides the advantages of the closed mold technique, such as

- Greater hole depth/length ratio
- Faster cycles
- Improved molded density
- Ability to mold undercuts with side draws
- Greatly reduced flash
- Increased numbers of cavities per mold

The introduction of the automatic transfer molding press as illustrated by the Hull 99C (Chapter 3, Figure 10) and the Stokes screw transfer unit (Chapter 3, Figure 12) gave further impetus to the continuing evolution of the transfer molding technique.

Both processes provided the possibility to mold parts fully automatically while enjoying all the advantages described above. Cost reductions of a major magnitude became available to the thermoset industry.

4.4 START-UP MOLDING PARAMETERS (Compression and Transfer)

	Allyls	Aminos	Phenolics	Thermoset polyesters
Preheat temperature	200–230°F 93–110°C	200–225°F 93–107°C	235–250°F 113–121°C	170–190°F 77–88°C
Mold temperature	325–340°F 163–171°C	260–340°F 138–171°C	350–400°F 177–204°C	320–330°F 160–166°C
Mold pressure	C–2.5–3K psi	2–6K psi	2–5K psi	1.5-3K psi
	T–3.5–4K psi	2–6K psi	4–7K psi	1.5–2k psi

Molding parameters for the high-pressure epoxy compounds, similiar to the allyl, phenolic and thermoset polyester compounds, will coincide with those listed for the phenolics.

Suggested molding parameters for the low-pressure epoxy encapsulation compounds in the table that follows:

Typical Molding Parameters for Low Pressure Encapsulation Compounds

Clamping pressure	Full
Transfer pressure (actual pressure on the molding compound)	200–1000 psi 15–70 kg/cm²
Cure time	90–180 sec depending on molding temperature
Mold temperature	345–365°F (175–185°C)
Transfer time (time from transfer slow-down to end of transfer cycle)	10–24 sec depending on shot weight and number of cavities
Preheat temperature	176–194°F (80–90°C)
Postcure	6 hr at 345°F (175°C)

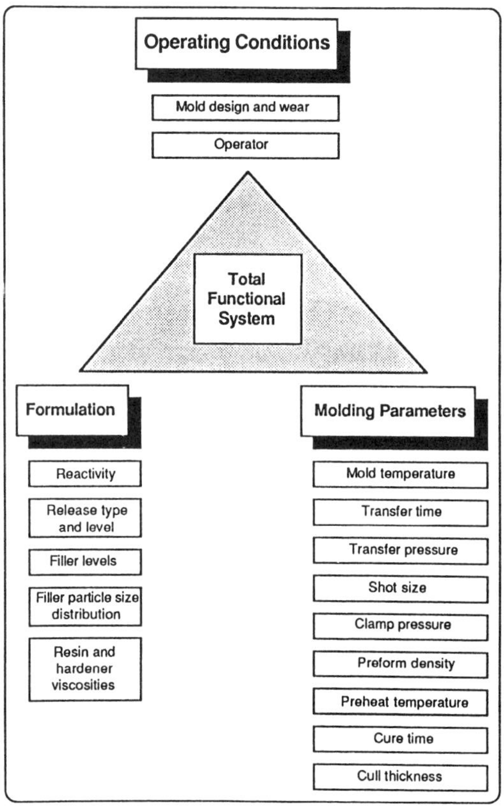

Figure 3. The thermoset modability of semiconductor molding compounds is a complex interaction of the formulation, the molding parameters, and the operating conditions.

Make certain that the transfer ram is hot. The transfer ram face should be 320°F (160°C) minimum before molding to prevent undercuring of the cull. Dielectrically preheat the compound to the recommended temperature (176–194°F, 80–90°C) for optimum results.

Mold cleaning is highly recommended using a melamine mold cleaner, followed by several shots using a mold conditioning compound that will lay down a light uniform coating or mold release on the cavity surfaces.

It is recommended that the molder be in contact with the compound supplier for advice on details as to mold cleaners or conditioners.

Suggested molding parameters for the silicone compounds are for the rigid and elastomeric compounds.

Rigid compounds are readily moldable in transfer molds with mold temperatures in the 300°F (149°C) range and transfer pressures in the 500–800 psi range. Cure times will be in the 1–3 min range. All of the parameters will be dependent on the molded part configuration with cross-sectional thickness determining the proper molding conditions (see Figure 3).

Elastomeric compounds are moldable in compression, transfer, and injection equipment. In all cases the molding pressures are in the 200–300 psi area with mold temperatures ranging from 260–340°F (127 to 171°C). Cure times will depend on the type of vulcanizing agent and the cross-sectional wall thickness of the molded article.

4.5 INJECTION MOLDING

The use of the injection molding process with thermosetting molding compounds had its origin in the early 1960s and today it would be safe to say that 60–75% of all thermoset molding is being done by this technique.

Molding compound suppliers have formulated their various compounds to provide sufficient "flow life" or gelation time at temperatures normally incurred in the barrel/screw area [230°F (110°C)–250°F (121°C)] so that precuring in the barrel rarely occurs. Also, molders have recognized that back pressure on the screw and screw rpms, when kept to the minimum—50 psi or less and 50 rpm or less—will keep the compound temperature in the desired range and offset the chance of precure.

These same compound formulations will exhibit very rapid curing as they pass through the sprue, runner, and gate areas on route to filling the cavity to maximum density. Generally, the injection time required to move the compound from the nozzle to complete cavity filling will be less than 10 sec and often less than 5 sec. The total processing time for a thermosetting molding compound—from nozzle to filled cavity—at 370°F (188°C) can be very low, with total cycles in the 17–20 sec range, depending on part configuration, mold design, and molding parameters (see Figure 4).[3]

	NON FILLS	POROSITY	STAINING	GAS BURNS	PARTS STICK	BLISTERS	WARPAGE	EXCESSIVE FLASH	EXCESSIVE SHRINK	ERRATIC INJECTION	NOZZLE CURE	SPRUE STICKING
Injection Pressure	↑	↑			↓	↑	↑	↓	↑	↑		↓
Back Pressure (Plasticizing)		↓	↓			↑			↑		↓	
Clamp Pressure								↑				
Material Feed Setting	↑	↑				↑		↓				
Nozzle Temperature		↓								↓	↓	↑
Barrel Temperature										↓	↓	
Mold Temperature	↓	↓	↓	↓	↓	↑			↓			↑
Injection Pressure Time							↑					
Injection Hold Pressure Time									↑			
Cure Time			↑		↑	↑	↑		↑			↑
Screw Plastication Delay											↑	
Material Flow	↑						↓	↓		↓	↓	
Polish Sprue Bushing												↑
Gates Size	↑					↑	↑					
Venting		↑	↑	↑		↑						
Injection Speed		↓		↓								
Polish Mold			↑		↑							

Trouble Shooting Guide

Rogers Engineering Thermosets are injection molded most successfully when press set-up parameters are within the ranges shown below. When adjusting press conditions to remedy a problem, it is recommended that only one change be made at a time, allowing sufficient time between adjustments for the effects to be determined.

Barrel Feed Zone Temperature	100°-160°F
Barrel Front Zone Temperature	140°-200°F
Nozzle Temperature	180°-210°F
Screw RPM's	40-80
Screw Back Pressure	25-75 PSI
Material Stock Temperature	220°-250°F
Injection Pressure	10-18,000 PSI
Injection Time (actual)	3-10 seconds
Injection Hold Pressure	3-10,000 PSI
Injection Hold Time	2-10 seconds
Mold Temperature	310°-380°F
Cure Time (1/8" section)	30-60 seconds

ROGERS

Rogers Corporation
Molding Materials Division
Box 550, Manchester, CT 06040
Phone (203) 646-5500
TWX 710-427-1836

Figure 4. Trouble shooting guide.

4.6 START-UP MOLDING PARAMETERS (Injection)

	Allyls	Phenolics	Thermoset polyesters
Stock temperature[a]	220–240°F 105–116°C	235–250°F 110–121°C	185–195°F 85–91°C
Mold temperature	320–340°F 160–171°C	350–400°F 177–205°C	320–330°F 160–166°C
Screw rpm	0–50 max	0–50 max	0–50 max
Back pressure	0–50 psi	0–50 psi	0–50 psi
Injection speed	3–10 sec	4–10 sec	As fast as possible
Injection pressure	3500 psi/min	4000 psi/min	3000 psi/min

[a]Compound temperature exiting the nozzle.

The above parameters are, of course, highly dependent on the flow properties of the molding compound, the part configuration, and the mold cavity layout, including runner and gate configurations, and are intended to be used as initial start-up guide lines. The lower pressure epoxy encapsulation compounds are not generally processed with the injection molding process due to the "mold open" delay while unloading and loading the insert frames.

Liquid silicone rubbers (LSR) are formulated with catalyst systems that provide high vulcanizing (curing) rates and can be injection molded at temperatures of 338–482°F (170–250°C), resulting in very rapid cycles. Injection pressures are in the lower ranges of 175–250 psi with curing times in the 30–60 sec range.

4.7 MOLDING QUALITY ASSURANCE

4.7.1 Incoming Raw Materials

- Molding compound
- Inserts
- External lubricants

Molding Compound

The most important raw material the molder is concerned with is the molding compound itself, and proper specifications covering those properties vital to their successful use must be understood and made clear to the supplier. Incoming quality control will be monitoring some of the following:

- Flow properties

  Cup closing (compression molding)
  Spiral flow (transfer)
  Brabender viscosity and time (injection)

- Cure rate
- Shrinkage
- Granulation—screen analysis
- Contamination—precure/color/moisture/content

Inserts

The use of molded-in metal inserts is a very common technique employed in thermoset molding, with the inserts generally purchased from outside vendors. Specifications governing the inserts are usually spelled out on part drawings that will list the following:

- Material
- Dimensions with appropriate tolerances
- Special requirements: plating, knurling, hardness
- Type of lubricant (often a water-soluble wax) for coating the exterior of the inserts to facilitate flash removal after molding

External Lubricants (Applied to Mold Surfaces)

- Contamination
- Purity

4.7.2 In Process

Moldings coming off the production line will be inspected on a regular and consistent basis for

- Key dimensions: The maintenance of key dimensions, flatness, and concentricity is often accomplished with use of cooling fixtures into which the molded part is securely held or clamped until it has reached room temperature. Parts removed from the hot mold will reflect stresses that need controlling until the part cools. Very often on long running, high-volume moldings involving critical dimensions, gauges are provided for speedy measurements on a regular basis.
- Molded density
- Cure
- Appearance
- Flash thickness—particularly on compression molded parts. Thickness exceeding 0.005 with general purpose moldings or 0.008 with impact type compounds will indicate:

  Lack of press clamping pressure
  Excess compound
  Compound flow too stiff
  Mold correction needed

4.7.3 Finished Goods

Final inspection will be based on those key elements that have been established by the customer's specifications and will generally be done on an acceptance procedure agreed on by the molder and the customer.

Statistical process control (SPC) methods have become a necessity with the advent, over the past decade, of more emphasis on product quality demands of the market place.

REFERENCES

1. Hunkar Laboratories Inc., Cincinnati, OH.
2. Hull Corp., Hatboro, PA.
3. Rogers Corp. MMD, Manchester, CT.

CHAPTER 5

POSTMOLDING OPERATIONS

5.1 FINISHING

It is very seldom that a molded article, ejected from the mold, will be suitable to ship directly to the customer without some additional work—regardless of the compound used or the molding process employed. It is true that certain compounds and certain molding processes do produce moldings that require less in the way of finishing operations, but, for the most part, there are inherent finishing needs that must be met on the vast majority of molded parts. In fact, there are occasions when the finishing costs will actually exceed the molding costs. Any time it becomes necessary to physically or mechanically handle a part an added cost presents itself. Thus it necessitates careful planning for these finishing operations in order that they be automated where possible to reduce those costs.

Post molding operations generally start with the removal of excess compound that is attached to the molding when it is ejected from the mold. This excess compound is referred to as "flash" and the quantity, thickness, and shape will depend on the molding compound used and the molding process employed.

5.1.1 Compression Molding: General Purpose Compounds

All compression moldings, regardless of the mold design, will produce "flash" on the entire periphery of the molded part along the parting line and in through molded holes. This flash will be 0.003–0.005 in. in thickness and often can be removed by the operator during the ensuing cure cycle. If the part is fairly simple in shape and contour, the flash will come off by wiping the parting line with gloved hands or by manual filing. If this is not possible to do within the cure cycle, the parts will be sent to an area where flash removal is accomplished with a tumbling or "shot blasting" operation. Tumbling may be done without any added media or, if there is need for a particular surface finish, the

Figure 1. Shot blaster.

tumbling operation can serve to remove the flash and also impart the desired surface.

Shot blasting (Figure 1[1]) is a technique wherein the molded part is subjected to a continuous "storm" or blast of a media with the operation done within a cabinet equipped with a continuous rubber belt that tumbles the parts as well as exposing them to the media blast, thus effectively removing all of the flash. There are two types of blasters, one operating as an ejector using compressed air in conjunction with the media and the other employing an impeller wheel with which large quantities of the media are hurled at the moldings at a rather low speed.

This principle is aimed at an overall coverage of the parts as opposed to concentrating the blast at specific areas. The impeller wheel method can be refined by supporting the wheel with a blower adding a stream of air that carries the media from the hopper through the suction tube to the impeller wheel.

Feeding the impeller wheel can also be accomplished with a bucket lifting device and an auger screw. The media will be fed to a reservoir from which it will fall freely into the impeller wheel via feed pipes.

The media used are of four varieties:

- Metallic
- Mineral
- Organic
- Synthetic (polymeric)

The organic type is used when deflashing plastic molded parts—walnut shells and cherry and olive pits being the most widely used.

The polymeric type consists of granules of thermoplastic compounds such as nylon, polyamide, and even some cured, granular, glass or mineral filled thermosets.

All of the shot blasting methods utilized with thermoset moldings require the addition of some moisture into the inside of the blasting cabinet to reduce the electrostatic charge resulting from the blasting operation. The moisture, in a fine spray, will minimize the adhesion of dust and media particles to the molded parts and to the interior of the cabinet.

Another version of the shot blasting utilizes a rotating table fitted with stations for retaining parts in a specific position. Each part passes under a preset blast of the proper type media while inside the cabinet, and the flash-free parts are automatically removed from the table on completion of the cycle.

5.1.2 Compression Molding: High–Strength, High–Bulk Factor Compounds

Once again, regardless of the compression mold design or molding process, there will be flash on the parting line and in the holes, but the flash will be 0.005–0.008 in. thick—very difficult to remove by tumbling or shot blasting. This necessitates the use of either manual filing or utilizing various types of belt or disc sanders or grinding wheels, or even trimming dies in trimming presses.

The choice of grit sizes coupled with proper speed settings is important to get the best and longest use out of the equipment. Heavy flash will require the coarser grit sizes and lighter flash will require the finer grit size. The use of special broaching tools will resolve the problem of flash removal from holes, whether on the surface or within the part itself. The design of the broaching tools is important in that the broach must not tear or splinter the molded part as it clears out the flash while passing through the hole.

Some of the locations and thicknesses of parting line flash that have to be removed are shown in Figure 2.

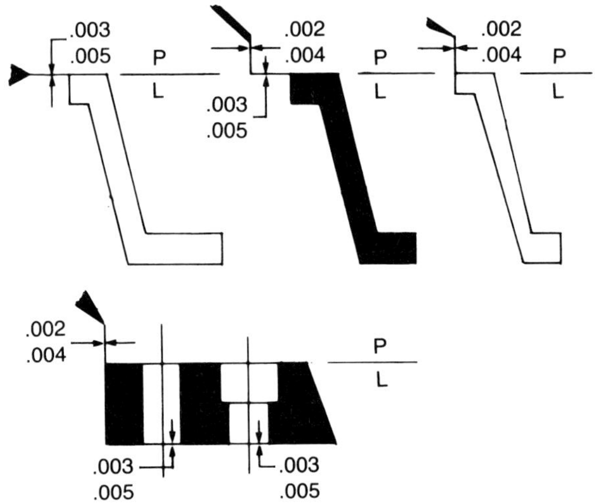

Figure 2. Types of parting/line flash.

5.1.3 Transfer and Injection Molding: General Purpose, Free Flowing Granular Compounds

The initial shot ejected from either a transfer or injection mold will possess a runner and gate system that must be removed to separate the parts—particularly if the mold contains more than one type of part or if there is a desire to keep the parts separated by cavity number. This operation is generally called "gate removal" and very often can be accomplished by the operator during the next cure cycle by simply snapping or cutting the part off the runner. There are mold designs that provide "tunnel gates" that will shear off as the part is ejected from the cavity.

The use of tunnel gating as shown in Figure 3 will produce a shot from the mold with the runner and gate becoming separated from the molded part by virtue of either utilizing a dual action K.O. system or through the cam or shear action of the K.O. pin pushing the runner and gate out of the mold. The dual action system will move either the molded part or the runner system out of the mold as an initial movement, thus accomplishing the separation.

A properly designed and constructed transfer or injection mold operating in the right size press with correct molding parameters carefully monitored will produce molded shots with little or no flash on the parting line or in holes. The core pins for molded holes can butt tightly against the inner mold surface and seal off any flash, or the core pin can telescope into the opposite face of the mold, again sealing off any entry of flash.

As shown in Figure 4, internally or externally threaded inserts can be molded flash free when the inserts and the molds are properly designed so as to seal

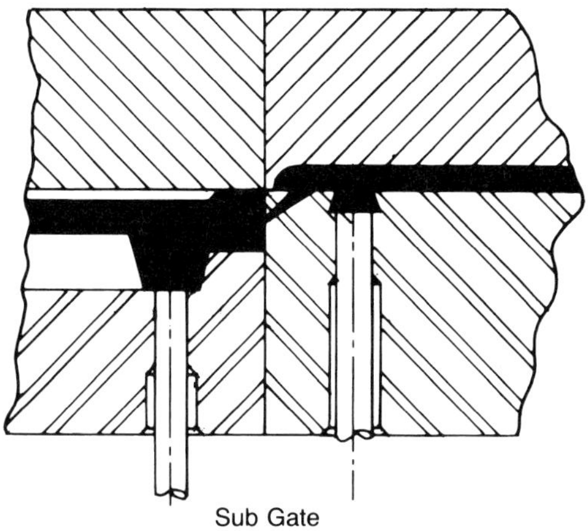

Figure 3. Tunnel gating.

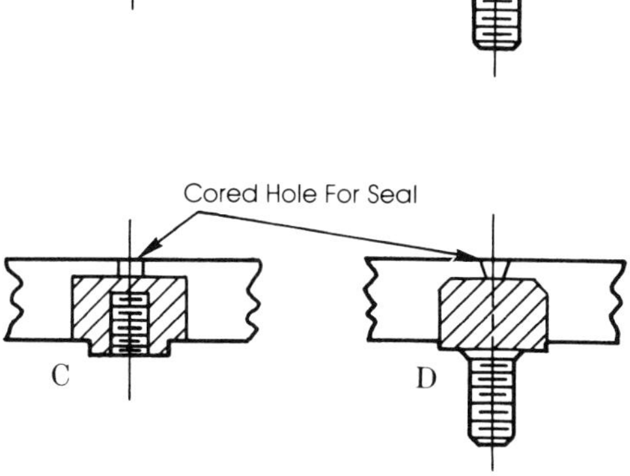

Figure 4. Molded-in insert types. In "A" and "B," a pocket in the mold will allow the insert to seat itself, and, with the mold closed, the insert threads are sealed off from the molding compound. With "C" and "D," a core pin on one side of the mold will provide the means to seat and seal the threaded portion of the insert.

the inserts off during the transfer or injection cycle. However, with certain types of inserts, such as flat, bladed electrical contacts, that cannot be sealed off by the mold and depend on the "fit" between the contact thickness and the slot in the mold, there is likely to be flash adhering to the contact requiring a removal operation. A multicavity mold producing a part with several contacts in each part will present quite a costly and time-consuming insert cleaning or flash removal operation, if done manually. By the utilization of a shot blaster with a traveling conveyor belt fitted with holding stations for retaining the parts in the proper position, this deflashing operation can be accomplished very economically. The performance of this can be enhanced by using a water-soluble wax coating on the inserts prior to molding. The parts are loaded into their stations at the entrance position and pass through a storm of blasting media en route to being unloaded flash free at the outlet position.

It is not unusual for transfer or injection molded parts to go directly from the press to shipping with only a quality control check made at the press.

5.1.4 Transfer and Injection Molding: High-Strength, High–Bulk Factor Compounds

These high bulk factor types of molding compounds do not lend themselves to the injection molding process due to their lack of pourability from the feed hopper of the press. The screw flights are often sufficient in depth and width to accept the compound, but the throat size of the hopper and the feed orifice to the screw are insufficient in size for the compound to pass through easily, causing the compound to "bridge" and not provide an even, consistent flow of compound to the screw. There are some innovative machine designs available as "special" designs with auxiliary equipment to force feed the compound to the screw. Also, there is a rather simple technique where the hopper is removed from the press and the operator manually feeds the screw directly with a preweighed amount of loose compound.

Transfer molding, with the use of preheated preforms, is ideal for molding these compounds with the only problem being the cost of manually made preforms. The normal preformer will have the same pourability problem from the material hopper to the preform cavity that the injection molding machine has, and the preforms have to be hand weighed and manually fed to a preforming device, not too unlike a compression molding press. Again, there are auxiliary devices available as "special" equipment that will auger weigh and feed these compounds to a preformer, but the accuracy of the preform weight is not equal to that of the manually weighed preforms.

It must be noted that a proper mold maintenance program is the key to keeping the molds in the proper condition where flash removal is concerned. A mold that has been damaged or has become worn in areas where the damage or the wear will cause the mold to flash excessively has to be corrected as soon as practical. Quality control, whose constant process overseeing can spot defective moldings and report the condition to the proper authorities, is one way to monitor this problem, but it is also incumbent on the molding operator and/or assistants to be aware of reporting such defects. The decision on the timing for mold corrections is one involving the proper authorities in production, sales, and management. Once a mold has either completed its production run or has to be repaired, the mold should be returned to the mold shop accompanied by a written report on its condition and with the last molding shot, taken from the mold, attached to it.

5.2 Postbaking

The vast majority of commercially molded thermoset parts meet the required standards as related to degree of cure and those parts have been serving their users very well. It is also recognized that those same parts have probably not been "fully cured" when subjected to a rigid examination of the state of their cure. This does not mean that those parts are defective in any way since they will be found to meet any and all electrical, mechanical, and thermal requirements called for by the procuring specifications. However, there are applications wherein the part specifications, due to certain environmental conditions or load constraints, call for properties in excess of the normal degree, and these applications will require the molded parts to be postbaked or exposed to high temperatures for a period of time so as to complete the cure to attain the optimum property profile.

Postbaking has been the subject of many studies over a number of years, mostly with phenolic molded parts using both single stage (resole) and two stage (hexa) compounds. All thermosets compounds will exhibit better property profiles when postbaked, but it is important to understand the effects of this added temperature cycle and the proper implementation of the process for optimum results.

Postbaking will accomplish the following:

- Complete the curing reaction
- Relieve molding stresses
- Drive off all volatiles incurred during molding
- Increase creep resistance
- Produce better retention of strength and modulus
- Improve dimensional stability
- Improve thermal shock resistance
- Raise the glass transition temperature (T_g)

The obvious benefits listed above can come about only if the postbaking procedure is done properly. Listed below are observations based on a recently completed study on the subject of postbaking:

1. The initial postbake temperature must be below the glass transition temperature (T_g) of the compound involved. Actually, a starting temperature below the actual molding temperature will suffice.
2. Parts of uneven cross-sectional thickness will exhibit uneven shrinkage.
3. The reinforcement system of the compound will, to a large extent, dictate the temperature/time cycles. Parts molded with compounds using organic reinforcements must be postbaked at lower temperatures than those molded with mineral or glass reinforcements.
4. Postbaking has to be done with a multistage temperature/time cycle as shown in the following sections.

5.2.1 Glass and Mineral Filled Phenolics

Single stage compounds:

1/8 in. or less—2 hr at 280°F (138°C)
4 hr at 330°F (166°C)
4 hr at 375°F (191°C)

Two stage compounds:

1/8 in. or less—2 hr at 300°F (149°C)
4 hr at 350°F (177°C)
4 hr at 375°F (191°C)

For any part exceeding 1/8 in. in thickness, it is recommended that the time be doubled for each additional 1/16 in. of thickness to obtain the best results. Longer time spans are preferable to increased temperature.

Note: Much of the above information on postbaking procedures and benefits has come from results of a number of studies conducted by the Rogers Corp., and the author is grateful to Bruce Fitts and David Daniels for sharing those results.

5.2.2 Allyl Compounds

Diallyl ortho phthalate: Seven hours at a range starting at 275°F (135°C) and raising the temperature 20°F/hr to 415°F (213°C).

Diallyl iso Phthalate: Eight hours at a range starting at 275°F (135°C) and raising the temperature 20°F/hr to 415°F (213°C).

5.2.3 Epoxy Compounds

Eight hours at a range starting at 275°F (135°C) and raising the temperature 20°F/hr to 415°F (213°C).

Note: In all cases, with all thermosets, it is recommended that the molded parts go directly from the hot mold to the baking oven and not be allowed to cool off prior to postbaking.

5.2.4 Thermoset Polyesters

Eight hours at 250°F (121°C) and raising the temperature 20°F/hr to a maximum of 350°F (177°C).

5.3 DECORATIVE/PAINTING/ PRINTING/PLATING

Decorative operations, such as overlay molding, are commonly used, particularly with melamine compounds. Articles such as dinnerware are compression molded with a very thin decorative melamine-impregnated paper overlay properly placed onto the molded article during a special mold opening, prior to final cure. The overlay is located on the part within the cavity and the mold reclosed, bonding the overlay to the surface of the molding. This technique has been primarily confined to melamine products where their broad color selection fits into decorative product lines.

Printing and painting operations are commonly applied to thermoset moldings and can be automated when the volume is sufficient to warrant the cost. It is always wise to consult with the paint and printing suppliers for their advice on the proper materials to use as well as their recommendations regarding part surface cleanliness necessary for painting or printing.

5.4 PACKAGING

Finished goods packaging will depend largely on

- Part size, configuration, and fragility
- Customer specifications
- Travel distance and mode of transport

Generally, moldings are shipped in cardboard cartons with appropriate liners with carton weights in the 50 lb range.

REFERENCES

1. Finmac Incorporated, Warminster, PA.
2. Rogers Corp. MMD, Manchester CT.

CHAPTER 6

MOLD DESIGN

6.1 COMPONENTS

Each molding process, compression, transfer, or injection is done with metal molds that are designed specifically for that process but, once again, there is a commonality in their basic design, construction, and structural components.

All thermoset mold bases are constructed with high-grade steels, generally hardened, polished, and hard chrome plated. The mold base, common to all molds, consists of

- Clamping plates
- Retainer plates
- Guide pins and bushings
- Return pins
- Support plates
- K.O. assembly
- K.O. box

Standard mold bases are available in a wide variety of sizes and they can be furnished with or without cavity "pockets" machined in the retainer sets to rough or finished dimension. Also available are other mold components such as

- K.O. pins—solid or sleeve type
- Support pillars
- Electric cartridge heaters
- Sprue bushings
- Transfer pots
- Rack and pinion gear sets
- Slide retainers
- Mold clamps
- Mold finishing and polishing supplies

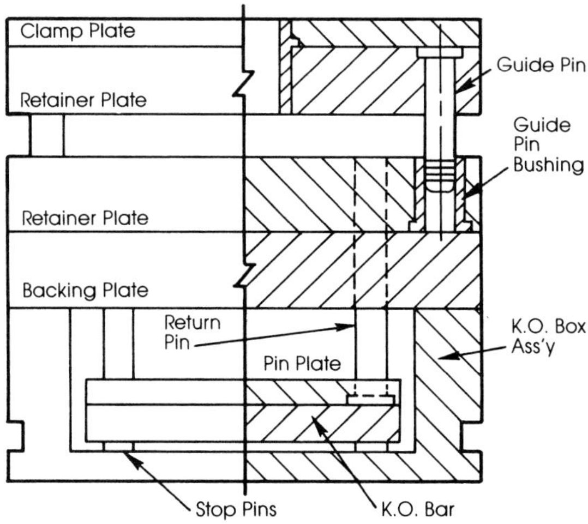

Figure 1. All thermoset molds have common components, starting with the mold base itself as shown here.

6.1.1 Mold Base Suppliers (U.S.A.)

D-M-E Co., 29111 Stephenson Hwy., Madison Hgts., MI 48071
Master Precision Molds, Inc., 866 Fairplains St., Greenville, MI 48838
Mold Base Industries Inc., 7501 Derry St., Harrisburg, PA 17111
National Tool & Mfg. Co., 100–124 N. 12th St., Kenilworth, NJ 07033

The mold, itself, is an extremely important unit in the thermoset molding procedure. A poorly designed mold, using inferior materials and lacking precision construction, will never produce quality moldings over any decent time span, whereas a properly designed mold, built with high-quality tool steels and with precision equipment by skilled craftsmen, will be capable of producing millions of moldings of superior quality.

Before the mold design actually commences there are a few important decisions to be made based on:

- Area, volume, and weight of the product to be molded
- Number of cavities needed to meet production needs
- Type of molding process
- Size of press needed
- Type of molding compound to be used
- Use of molded-in inserts—type and quantity to be used

The mold designer will receive the above information from the cost estimator, who will have developed it when preparing the cost estimate. During the mold design phase it is incumbent on the designer to keep in constant contact with the product designer, molder, and compound supplier to clarify any problem areas that might arise prior to mold construction. This is the time to make any revisions and not after the mold has been placed with an outside mold maker or with an internal mold shop.

6.1.2 Molding Draft Requirements

Regardless of the mold type, compression, transfer, or injection, all molding surfaces parallel to the mold opening movement must incorporate provision for easy release of the molded part from the mold on ejection. This is accomplished by providing these surfaces with a draft angle of ½^{1} minimum. This draft angle must be included in any compound shrinkage calculations, especially on very close tolerance dimensions.

Note: See Standard Draft Angle chart on page 188.

6.1.3 Molding Shrinkage Calculations

All thermoset molding compounds shrink as they come out of the mold and cool off. This shrinkage factor is given in the data sheet values for any specific compound with attention given to the rheology of the compound. The resin, itself, does the shrinking, but the reinforcement and resin ratio as well as the molding process will greatly influence this phenomenon.

A soft flow, general purpose phenolic molding compound will shrink about 0.006–0.008 in./in., whereas a stiff flow, general purpose phenolic compound will shrink about 0.005–0.007 in./in.

Generally speaking, all higher strength compounds will shrink less than lower strength compounds, even with the same amount of resin in them.

It is always wise to consult with the molder and the compound supplier prior to finalizing the mold design, providing both with as much information regarding the compound, the mold design, and the molding process as possible.

6.2 COMPRESSION MOLDS

There are four basic compression mold designs that will now be looked at in detail:

- Flash type
- Positive type
- Semipositive type
- Semipositive subcavity gang type

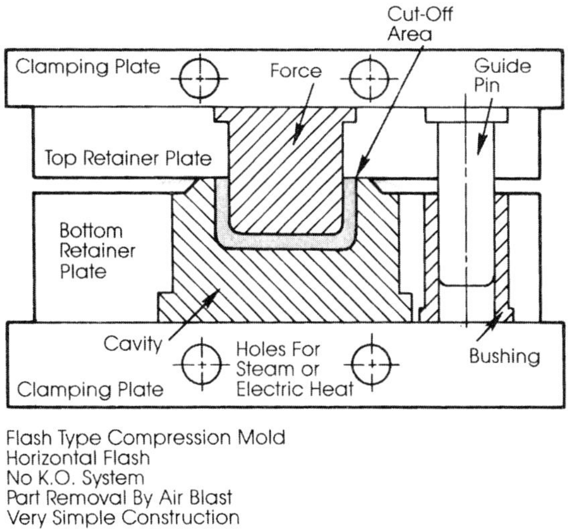

Figure 2. Flush-type mold.

6.2.1 Flash Type

Flash type (Figure 2) is the simplest of all the thermoset molds and is still widely used in the production of simple shapes such as molded dinnerware, ashtrays, closures, or knobs. The mold will be constructed with a retainer set for the force and cavity blocks and providing guide pins and bushings for positive mold alignment when opened and closed. Holes for heaters are drilled into the clamping plates. The force and cavity should be built with a good tool steel, through hardened to resist the compression load on the cut-off lands of the cavity. Both units will be highly polished and hard chrome plated for easy mold release and a superior surface finish. A relatively simple shape such as this will not need a K.O. system and can be released from the cavity with an air blast.

6.2.2 Positive Type

As seen in Figure 3, the force in the upper mold half is allowed to enter the cavity in the lower half with a tight fit between the two components (0.002–0.003 in. clearance), providing direct and positive pressure on the compound in the mold cavity as well as correct alignment of the force and cavity during mold closing. As in all compression molding, the mold charge will be in

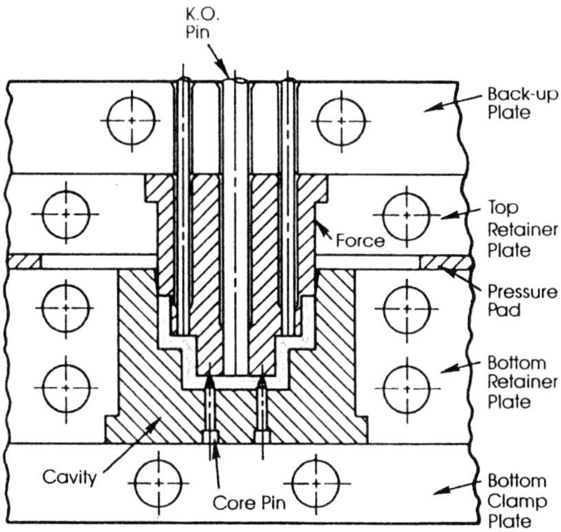

Figure 3. Fully positive compression mold.

excess of the part weight and the excess compound will be allowed to escape in a vertical direction and under controlled conditions by the use of "bleed outs" or flats that are ground 0.010–0.015 in. deep in strategic locations on the side of the force that enters the cavity (Figure 4).

Another virtue of this type of design is that the excess compound, commonly called "flash," is vertical and easily removed by manually pressing the upper face of the part against an abrasive surface and rotating it until the flash is entirely gone. The positive molding will promote excellent molded density while the hardened pressure pads at the parting line of the mold will serve to prevent damage to the mold in the event that the mold is closed with no compound in the cavity.

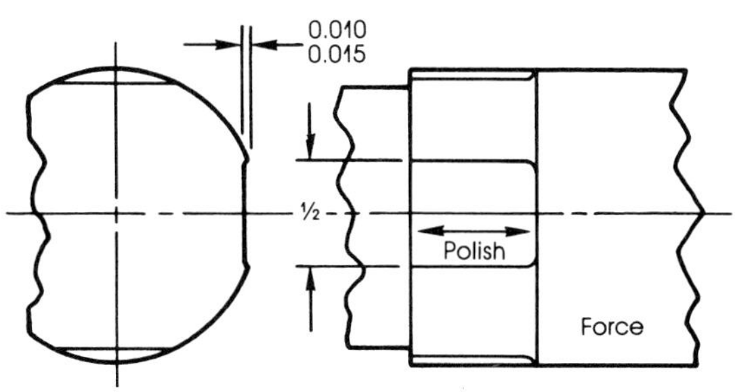

Figure 4. Bleedouts must be well polished in the direction of the arrow.

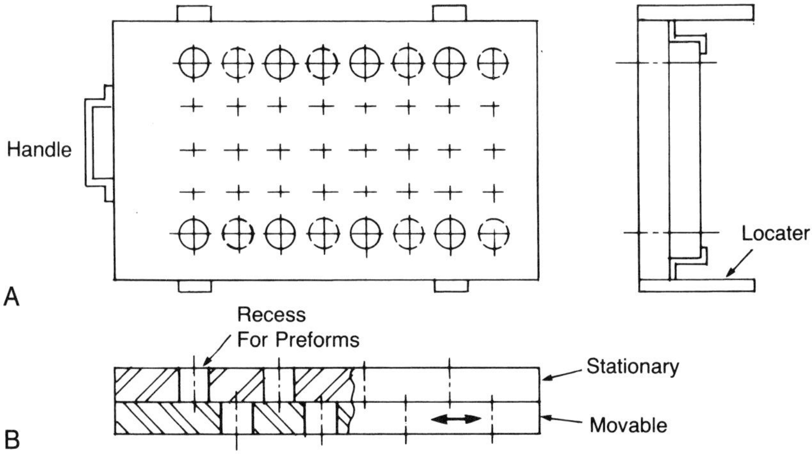

Figure 5. Loading board for 20-cavity mold.

This type of mold, often multicavitied, is generally run semiautomatically, using preheated preforms that are fed into each cavity manually or with the use of a "loading board" (Figure 5) that will feed all cavities at one time, thus reducing the mold open time and cycle time.

The preforms are placed in the holes in plate (A), which are in alignment with the mold cavities, this loading being done during the previous molding cycle. With the mold open and ready to receive the next load of preforms, the board is properly located over the cavities and plate (B) is moved to bring its holes directly under those of plate (A), allowing the preforms to fall into the cavities. There are loading boards designed that will be capable of being placed within a dielectric preheater so the preforms will be of the proper temperature when fed to the mold.

6.2.3 Semipositive Type

The semipositive mold (Figure 6) resembles the positive mold except that the force actually "lands" within the cavity well, producing a horizontal as well as a vertical flash. The effect of landing the force provides a better control of the flash thickness as well as the part dimensions that are determined from the parting line of the mold. Again, hardened pressure pads are mounted on the mold at the parting line to absorb the high molding pressure and prevent damage to the land areas from the constant opening and closing of the mold when in production.

In both positive and semipositive molding, the telescoping of the force into the cavity serves the important function of maintaining precise mold alignment and will produce parts with superior molded density.

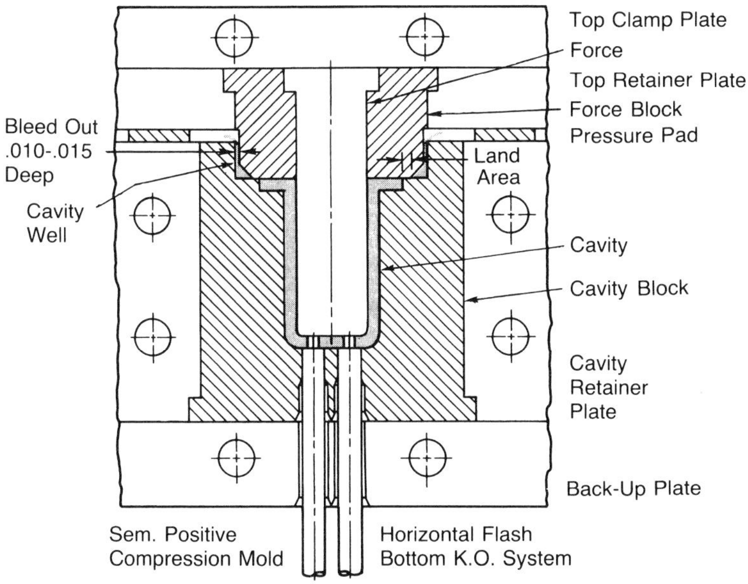

Figure 6. Landed semipositive mold.

6.2.4 Semipositive Gang Type

This mold design (Figure 7) utilizes the benefits derived from the positive and semipositive designs along with the ability to load the compound into a single, centrally located area instead of having to feed each individual cavity. The cavities are mounted into a hardened retainer "shoe," as shown, with the compound being fed into the cavity area and the mold closed under pressure. The compound will flow into each separate cavity and the excess compound will serve to connect all the cavities at the parting line. The flash adjacent to the perimeter of the molded parts will be 0.003–0.005 in. in thickness whereas the remaining compound will be about 0.015 in. thick, keeping the entire shot intact for removal from the mold. The parts are separated by the operator during subsequent cure time or the shots are sent to the finishing room where they are mechanically separated in a secondary operation such as tumbling or wheelabrating.

As can be seen from the various illustrations, the compression molding process requires high molding pressures, and it is imperative that the mold components be of high grade tool steels and the mold built with ample support throughout to prevent any plate deflection over the life of the mold. AISI-SAE P20 steel, when properly heat treated, will provide excellent resistance to high molding pressures, and is capable of being highly polished and hard chrome plated.

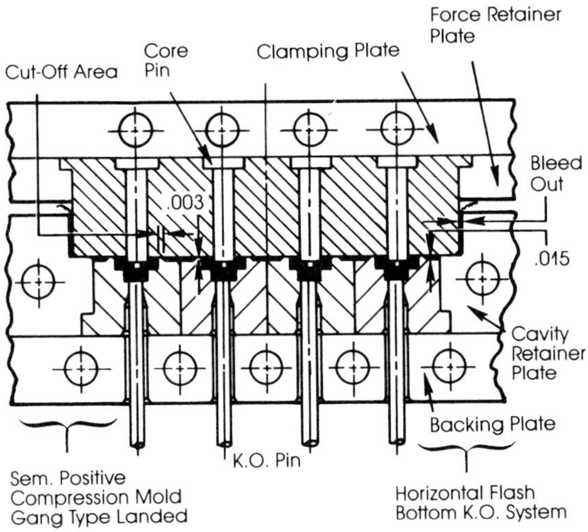

Figure 7. Semigang mold design.

The high pressures, noted above, coupled with stiff flow compounds, make it very difficult to mold holes in the part that are greater than 1:4 in ratio of diameter to length (Figure 8), i.e.,

- 0.062 in. diameter – ¼ in. length
- 0.125 diameter – ½ in. length
- 0.250 diameter – 1 in. length

These same pressure/compound conditions will also prove damaging to any fragile force or cavity sections, and such sections, should be constructed as remov-

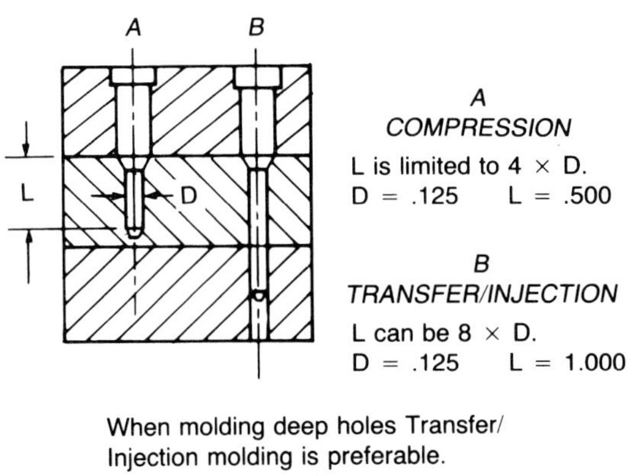

Figure 8. Core pin length/diameter ratio.

able inserts that can be easily replaced so that an entire new cavity or force will not have to be made in the event of such damage.

6.3 TRANSFER TYPES

6.3.1 Pot Type Transfer Mold

The original "closed" mold design utilized the compression molding press to supply both the mold closing pressure as well as the necessary pressure to force the compound out of the pot through the sprue hole into the cavity (see Chapter 3, Figure 8). The mold design provided a central loading area, the ability to mold parts with a minimum of parting line flash and flash in molded holes. Because the mold was never completely closed until all of the compound had entered the cavity, it was difficult to obtain parts that were flash-free, and molded-in inserts could not be entirely sealed off. Another problem area was the amount of waste compound in the pot cull, sprue, and runner system. The area of the pot had to be greater than the combined area of the entire molded surface to keep the mold closed when under pressure.

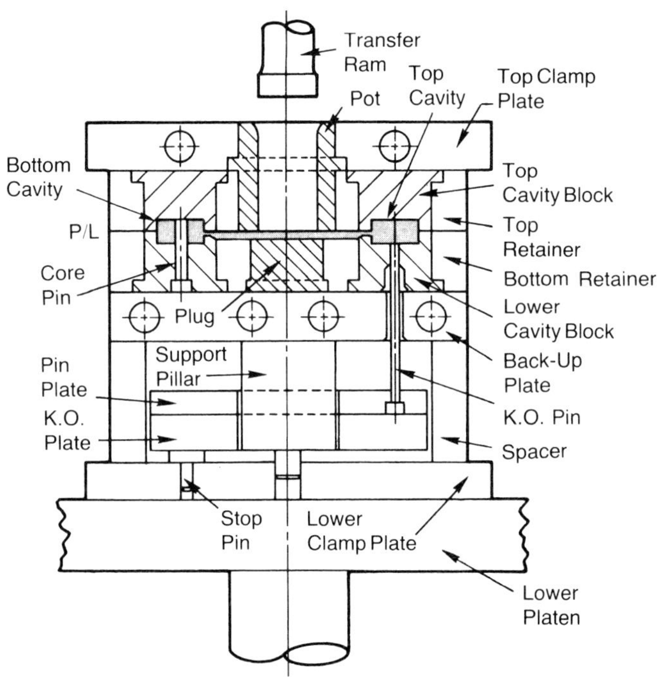

Figure 9. Plunger/transfer mold.

6.3.2 Plunger/Transfer Type

The plunger type transfer mold (Figure 9) is a closed mold using a centrally located pot into which the molding compound is fed with the mold in a closed position. The transfer ram, acting separately from the clamping ram, will force the compound, under pressure, out of the pot into a runner and gate system feeding the individual cavities of the mold. The lower, main clamping ram serves to keep the mold tightly shut during the entire molding cycle and also provides sufficient pressure to open the mold once the cure time has elapsed. With the molding pressure concentrated on the end of the transfer ram as it enters the center of the mold it becomes necessary to insert a hardened steel support pillar between the backing plate and the lower clamping plate to prevent any deflection of the mold during molding. The molding areas of the mold, consisting of the upper and lower cavity faces, runner blocks, and lower mold "plug," should be mounted into their retainer plates so that the molding faces are projecting above the face of the retainer plates by about 1/32 in. This will reduce the area of the mold under pressure, making the available clamping pressure more effective.

6.3.3 Hot Cone Transfer Type

The concept of utilizing the shear effect when the compound passes over the hot mold surface as it enters the mold can also be taken advantage of in transfer molding as shown above (Figure 10). The cone shaped recess in the end of the plunger will encompass the similiarly shaped, heated core pin leaving a clearance of 1/16th–3/32 in., allowing the compound to flow over the core pin and through the clearance to enter the runner system. Again, the temperature increase of the compound resulting from the shear action will reduce the vis-

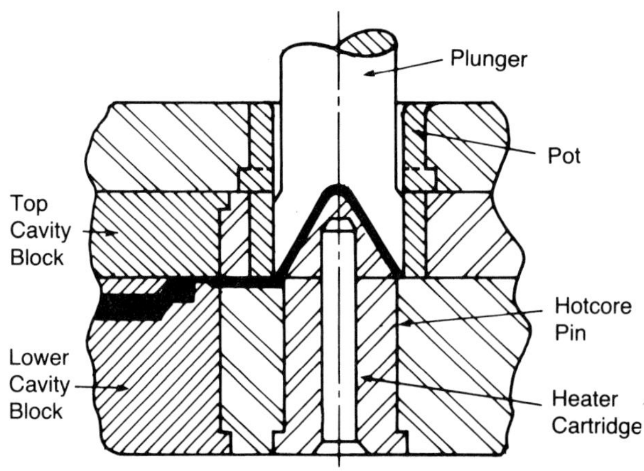

Figure 10. Hot cone transfer.

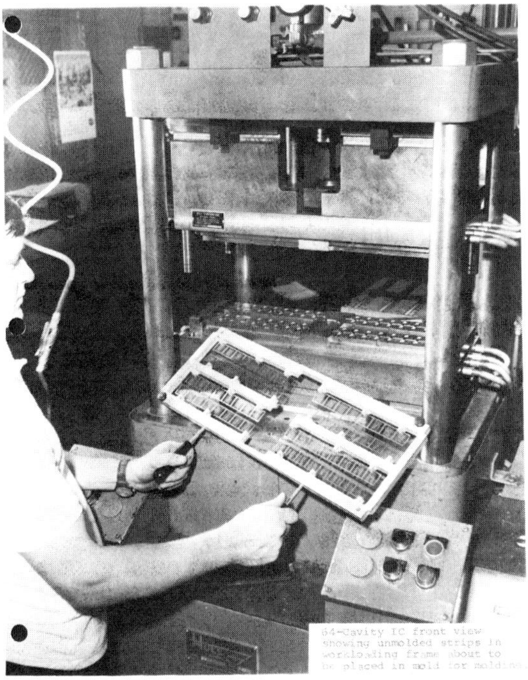

Figure 11. 64-Cavity IC front view showing unmolded strips in workloading frame about to be placed in mold for molding.

cosity of the compound and also result in more rapid curing and speedier molding cycles.

6.3.4 Encapsulation Molds

6.3.4.1 Standard Plunger/Transfer Mold

The plunger/transfer mold as applied to the encapsulation molding process is basically a closed mold that is designed to accommodate an insert loading device as shown in Figure 11.[1] The loader is positioned in the mold, the mold closed, and the preheated compound fed into the pot in the upper mold half. Two loading boards are generally employed, with one board in the mold and the other board being unloaded and reloaded outside the mold. The number of cavities in a single mold can be in the hundreds, depending on the production needs and the ability to fill out all the cavities to a proper molded density.

Figure 12[2] illustrates a 1200 cavity plunger/transfer encapsulation mold—currently the world's largest—which incorporates five molding operations in a single unit: specially developed split loading frames with a unique cull breaker design and pin positioning both top and bottom.

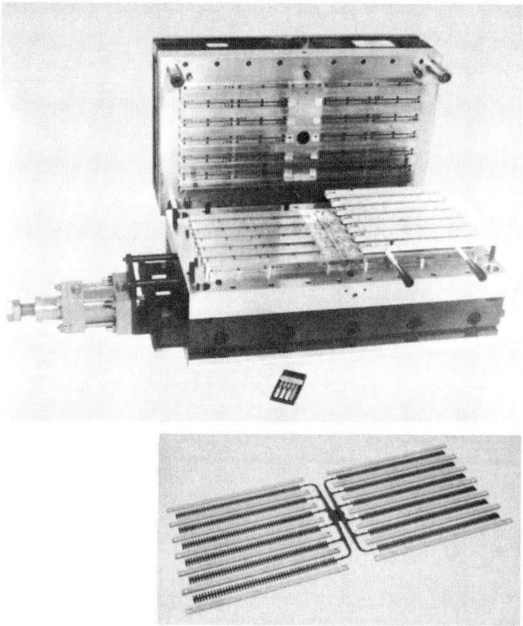

Figure 12. 1,200-Cavity plunger transfer encapsulation mold.

6.3.4.2 Aperture Plate Mold

The aperture plate mold (Figure 13[2]) design utilizes a new concept of insert loading that has the following features:

- Zero flash
- No ejector pins
- Positive package alignment
- Ability to mold various products without having to build complete molds for each product
- Reduced compound usage

6.3.4.3 Multiplunger Molds[3]

The multiplunger transfer molds differ from the conventional plunger/transfer mold in that they utilize more than one pot/plunger, enabling the molder to mold a number of different devices in the same mold or mold the same device with shorter runner systems. The conventional system employing several hundred cavities will have very long runner systems and is quite adequate for the

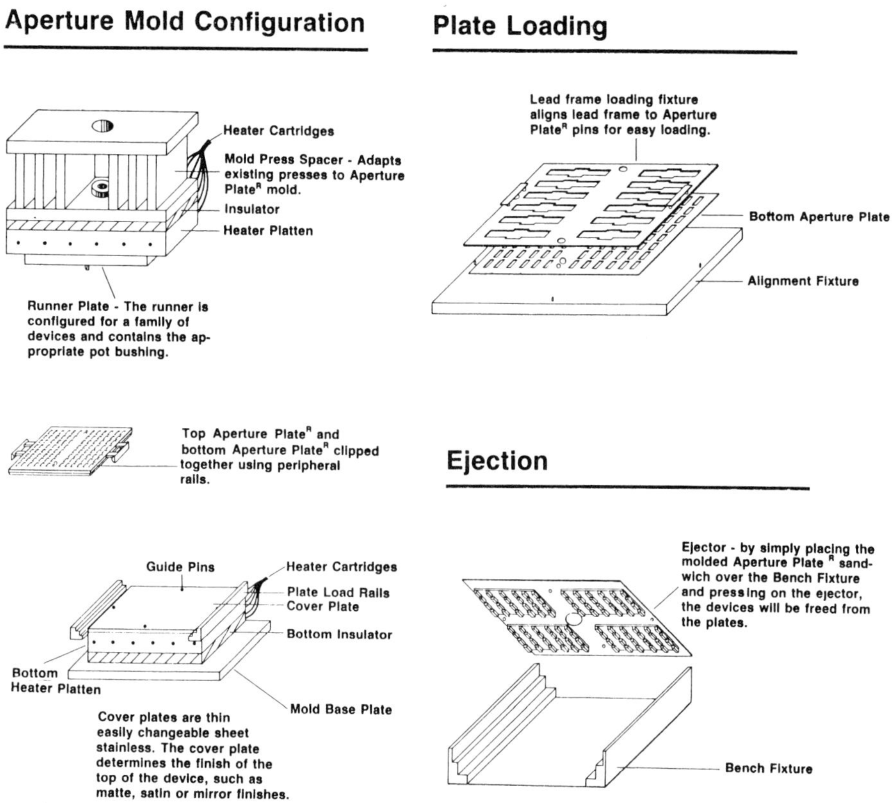

Figure 13. Aperture plate mold.

mass production of certain less-costly electronic devices. However, there are other devices that are very costly to manufacture and must be molded in molds that will ensure a very minimum defect rate, while still being capable of high productivity. The use of the multiplunger system has proven able to meet this demand.

6.4 SCREW TRANSFER MOLDS

The screw transfer mold design, as illustrated in Figure 13 of Chapter 2, incorporates a centrally located pot in the lower half of the mold and positioned so as to line up with the nozzle of the horizontal extruder unit as shown above. The transfer ram works in an upward movement, delivering the heated compound to the center of the mold for distribution through the runner and gate system to the mold cavities. The K.O. system is located in the upper mold half with the entire molded shot ejected onto a pick-up tray that moves across the face of the mold as the mold opens. The tray, with the molded shot, then retracts from the mold and deposits the shot onto the operator's bench or onto a moving conveyor belt.

The hot cone mold design is also possible with the screw transfer process.

6.4.1 In-Line Screw Injection Molds

Mold designs for screw injection molding are very similar to those used for transfer or screw transfer molding in that the molds are in the closed position when the heated compound is forced, under pressure, into the mold. However, the centrally located pot, used in transfer molds, is replaced with a "sprue bushing," which is contacted by the molding machine nozzle and through which the compound flows into the runner system. The molds are constructed with clamping plates, retainer plates, backing plates, support pillars, and a K.O. system. All molding components, such as cavities, runner blocks, sprue bushing, and slide plates, are built with high grade tool steel, hardened, highly polished, and hard chrome plated for wear resistance, good mold release, and superior molded part surface finish.

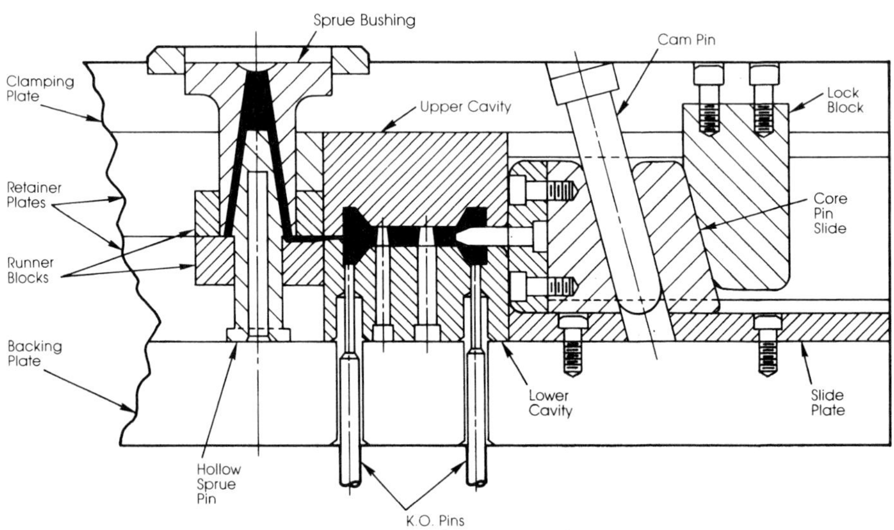

Figure 14. Side draw screw injection mold.

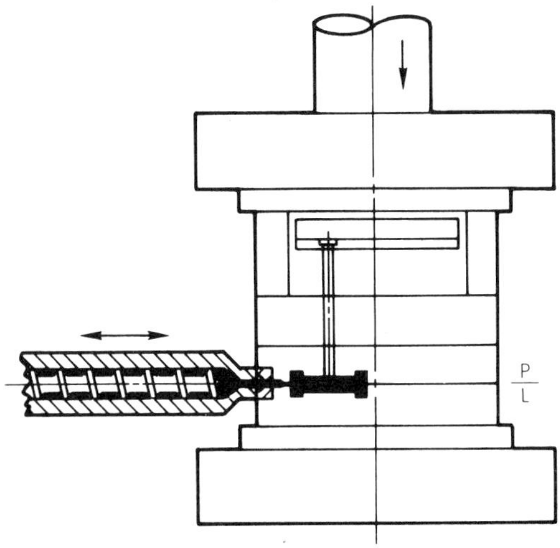

Figure 15. Screw injection parting line.

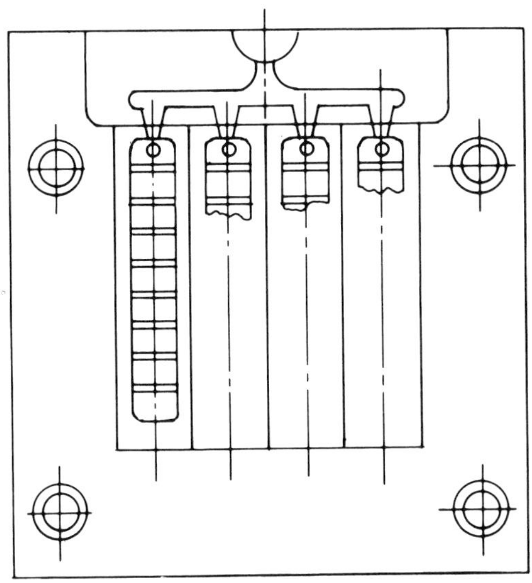

Figure 16. Parting/line injection mold.

6.4.2 Side Action Molds

With closed molds such as transfer or injection molds, the ability to mold holes or undercuts that are not "in-line" with the direction of the mold opening becomes simpler since it is possible to utilize the mold opening force to provide the means to operate cam actions for this purpose. Figure 14 Illustrates a cam action mold.

6.4.3 Parting Line Injection Molds

Injection molds (Figures 15 and 16) designed for use in parting line molding presses require a two-piece sprue bushing as illustrated in Figure 17. The value of the P/L mold design lies in the ability to load inserts with the mold face being on a horizontal plane, eliminating the danger of insert dislodgement when the mold is closed. Another benefit is that very long parts can be end-gated with shorter runner systems.

6.5 RUNNERLESS INJECTION MOLDING

The Durez Resin Div. has also patented a mold design system that is described as runnerless injection/compression molding (RIC*) that combines these two molding processes in a manner that eliminates the waste compound of cured sprue and runners and, at the same time, provides the beneficial effects of compression molding on the molded density of the final molded part. This is accomplished with the initial injection of the heated compound into a partially closed mold cavity. With the cavity filled with compound, the mold is completely closed, compressing the compound to maximum density.

The RIC process consists of three types, all of which produce compression molded parts with no runner or sprue waste:

- Direct sprue gated (DSG): DSG injects the heated compound directly into a partially closed single cavity. The mold is then closed, as in compression molding.
- Common Pocket (CP): CP injects the heated compound into a partially closed common pot, or loading well, of a subcavity compression mold. This process provides for multicavity designs.
- Manifold: The manifold process injects the heated compound through a centrally located manifold system, measuring and directing the correct amount of compound into each of several cavities. The cavities can be alike or dissimiliar in weight or configuration and the moldings are all runnerless.

There are added advantages to the RIC process, such as

- Reduced cycle time
- Lower compound usage and waste
- No gate design
- Suitable for standard compounds
- Vertical parting line

The RIC process and equipment are covered by U.S Patents 4,238,181; 4,260,359, and 4,290,744 as well as pending patent applications 6.10.

6.6 HOLLOW SPRUE/SPRUELESS MOLDS

In the ongoing evolution of thermoset injection molding techniques, it became apparent that the sprue, itself, was often the thickest portion of the entire molding and that, as a result, it determined the curing cycle. By "coring" out the sprue with a heated, tapered core pin, the sprue was no longer controlling the molding cycle.

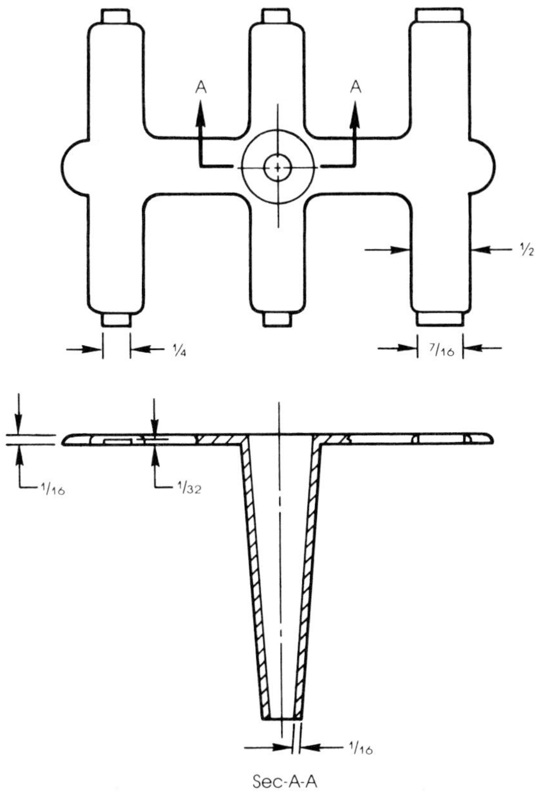

Figure 17. "Min/max" hollow sprue molding system.

By combining the "hollow" sprue with a flat, thin runner it became possible to take advantage of the shear heat generation caused by the passage of the molding compound over the hot molding surfaces to the sprue area, runner, and gate system and on into the cavity itself. The resulting elevation of compound temperature produced the beneficial effects of reduced viscosity and accelerated cure rate. In all, molding pressures are reduced and very rapid molding cycles result. The process is referred to as the "min/max" hollow sprue molding system (Figure 17).

General Electric Co. held some basic patents on what was described as "hypothermal" molding involving much of above technology and when G.E. departed the thermosetting compound field, the patents were sold to the Durez Resin Div. of Occidental Chemical.

6.7 SHORT HOLLOW SPRUE/EXTENDED NOZZLE DESIGNS

Success with the longer hollow sprue led to a desire either to shorten the sprue length or eliminate it completely. To shorten the sprue requires a new sprue bushing design in which the bushing is jacketed to allow circulation of coolant to control the compound temperature at the same temperature range as had been accomplished in the nozzle of the molding machine. The design has been successfully implemented, but its broader acceptance has been limited because it requires that each mold have its own special sprue bushing (Figure 18).

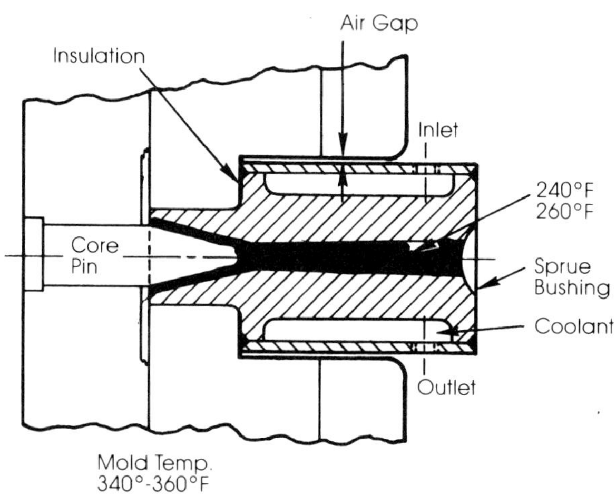

Figure 18. Short hollow sprue molding system.

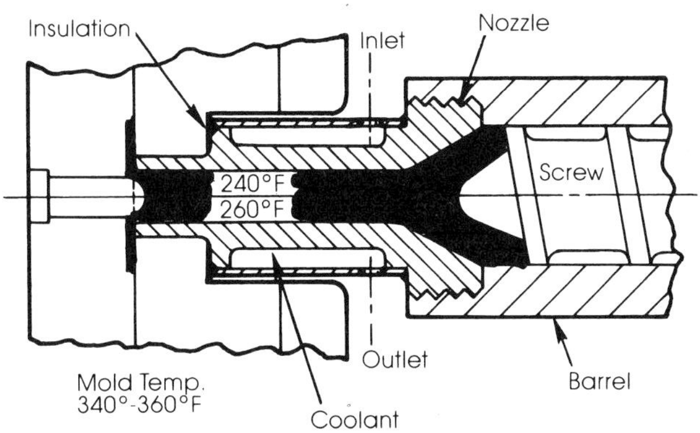

Figure 19. Extended nozzle.

The jacketed sprue concept was redesigned to become an integral component of the molding machine in the form of an extended nozzle that could replace the normal nozzle and would fit all molds (Figure 19). It actually eliminated the mold sprue bushing entirely, with the extended nozzle entering the mold and extending to the parting line of the mold.

Combining the extended nozzle with the flat, thin runner and gate system has succeeded in

- Less compound use and waste
- Reduced finishing costs
- Sharply reduced molding cycles

6.8 MOLD VENTING

All closed molds, transfer, screw transfer, or injection type, have need of "venting" on the parting line or in deep mold crevices to allow the hot air resident in the cavities or the volatiles generated by the compound while entering the mold to escape.

The usual technique for venting (Figure 20) is to grind "flats" (0.003 in. deep × 0.250 in. wide) on the face or land of the cavity in locations where the cavity is last to fill. The movement of the compound within the cavity tends to force the volatiles or hot air out of the cavity through the flats or vents. Sometimes it is possible to predict the flow pattern as the mold design becomes complete and the vent locations can be located and put into the mold during the final assembly. Otherwise the vent locations may come after the first shots have been taken and the flow pattern has been established. Lack of proper venting will be visible in surface porosity, actual burn marks, and mold staining in areas where the volatiles have not been able to exit the mold.

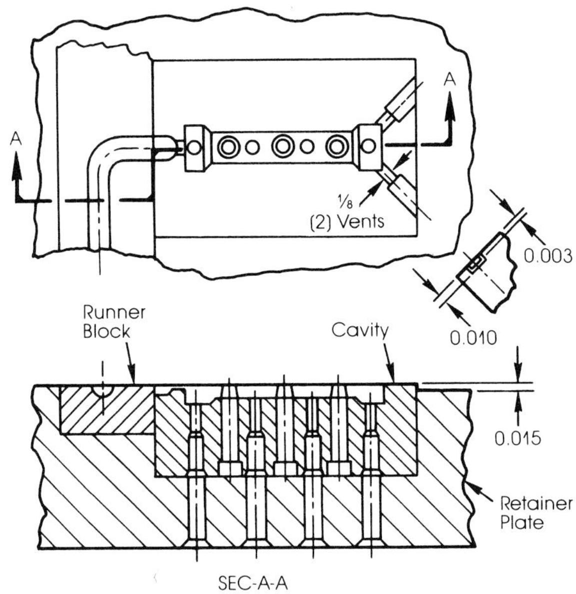

Figure 20. Venting technique.

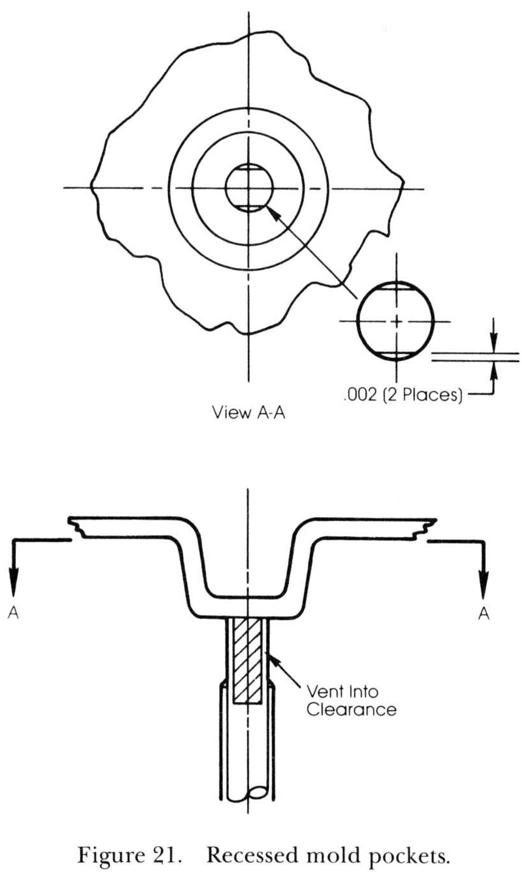

Figure 21. Recessed mold pockets.

Figure 22. Mold venting system (courtesy Hull Corp.).

Recessed mold pockets or deep sections that are difficult to vent can be aided by the use of K.O. pins with "flats" or "bleed-outs" ground on their sides that will allow the volatiles to escape (Figure 21).

When normal venting techniques are inadequate to prevent the entrapment of either hot air or other volatiles, vacuum venting should be considered (Figure 22).[4]

The following is from John L. Hull's Paper, the "Principles of Vacuum Venting."*

When normal cavity venting techniques for closed mold molding (transfer or injection, with thermostat or thermoplastic materials) are inadequate to prevent air entrapment, vacuum venting should be considered.

What Problems Arise from Air Entrapment?

Entrapped air builds up to the same pressure as the molding material, finally stopping further cavity fill, resulting in voids in the molded part. In addition, the high pressures experienced by the constantly shrinking volume—and increasing pressure—lead to increased temperatures of the entrapped air, according to Charles Law, resulting in temperatures so high that the plastic may, in fact, char. With thermosetting plastics, such heat accelerates cure of the formulation, stopping fluid flow and resulting in precure of the material, generally before cavity fill.

*Courtesy Hull Corporation, Hatboro, PA.

What is Vacuum Venting, and What is the Theory Behind It?

Vacuum venting is a process whereby the cavity of a mold is exposed to vacuum prior to and during the flow of material into the cavity. Theoretically, if the vacuum in the cavity is "perfect"—that is, if the vacuum has removed every last molecule of air before the material flows into the cavity (and assuming a "perfectly" sealed cavity), then there can be no—absolutely no—air entrapment in the cavity, even if the vents are closed by material prior to the complete fill.

In practice, however, there can be no "perfect vacuum." But if most of the air is removed, then the small amount of remaining air, even if eventually trapped in an unvented portion of the cavity, may be inadequate to prevent complete fill. With many thermoset materials, it has been demonstrated that under pressures commonly encountered in transfer or injection molding, small amounts of air may in fact be taken into solution—fully absorbed into the molding material—leaving no void in the finished molded part.

In using vacuum venting, therefore, it is important to create the lowest possible air pressure in the cavity as the material flows in through the gate, often at high velocity. A vacuum of 25″ Hg. is far far better than 20″ Hg., and 28″ Hg. is considerably better than 25″ Hg.

An exception to the above credo is when the vent or vents can be located precisely where the final cavity fill occurs. In such rare instances, vacuum will speed the air escape from the cavity and will be effective up to the final second of cavity fill, even if only a modest vacuum is achieved.

But in most instances, vents can't be so fortuitously positioned that they are at the precise point of final cavity fill.

How Does One Achieve the Ultimate Vacuum in the Cavity?

In principle, in any vacuum system, the vacuum pump does not "suck" the air from an enclosed space. Instead, the higher pressure air in the cavity tends to flow toward the lower pressure at the vacuum pump. Flow of air is restricted by small passages, by changes in direction, and by distance of flow. Even with a vacuum of, say, 30″ Hg. at the vacuum pump, vacuum in a mold cavity only 3′ distant may be only 10″ or 15″ Hg.—because of the restrictions of vents, flow space outside cavity vents to the vacuum port(s) in the mold, and restricted vacuum lines from vacuum ports to the pump.

It is vital, therefore, if one wishes to insure maximum benefit from vacuum venting, that the air flow restriction between cavity and vacuum pump be kept to a minimum.

One more point must be recognized: when a vacuum pump is "digesting" a considerable flow of air, as, for example, when a cavity at atmospheric pressure is to be quickly evacuated, the vacuum pump momentarily ceases to pump at, say, 28″ Hg., but may be pumping at, say, 15″ Hg., until the bulk of the air has passed through the pump.

Because of the factors outlined above, a good vacuum venting system must include the following features:

1. A vacuum pump, capable of operating at a blank-off pressure of 100 microns or less, and with a through put capacity of 10 or more cfm.
2. A "vacuum reservoir," of at least one fourth the pumping speed per minute of the pump. That is, if the vacuum pump is 10 cfm, the reservoir should be at least 20 gallon capacity (20 gallon = about 2.5 cubic feet).
3. A solenoid operated valve should be located between the reservoir and the mold. This valve is closed as soon as each cavity is filled, and opened before cavity fill begins, to enable the vacuum reservoir to be evacuated to minimum pressure while the mold is open for part removal, cleaning, and reclosing. (In transfer molding, the valve should be opened as soon as plunger enters the pot. In injection molding, the valve should be opened as soon as the nozzle meets the sprue bushing, and a 3 second delay should be programmed before the injection boost starts, thereby giving more time for initial cavity evacuation before material enters the cavity.
4. All vacuum lines, between the vacuum pump and the reservoir, between the reservoir and the solenoid vacuum valve, between the valve and the mold, and between the mold vacuum port and the cavities, must be as large as practical, and as short as practical. We strongly recommend 3/4″ I.D. lines, valve openings, and fitting I.D.'s, for most molding operations.
5. Special attention must be paid to the air flow from mold cavity to the mold vacuum ports.
   a) We recommend several wide vents at the parting line—as wide as 1″, with vent depths of perhaps 0.003″ for a distance for no more than 0.125″ from the cavity, then deepened to about 0.006″ for no more than another 0.375″.
   b) Beyond that distance (usually the land area of the cavity), the die base should be hogged out to allow clearance between top and bottom parting surfaces of at least 1/4″, with width of flow path never less than 2″—in effect giving a rectangular flow path equivalent to 1/2 sq. inches.
   c) From this "manifold," where air escaping from cavities is flowing, the opening to the outside surface of the mold base should be a minimum of 1/2″ I.D. (preferably 3/4″), and with as few changes in direction as practical.
6. Cavity vents must be smooth, polished hard chrome plated, to insure no stickage of material in the vent from one shot to the next. Ideally, any material which flows into a vent will come out with the molded part, and be thin enough (0.003″) to break off easily after cooling, with no significant vent scar. Mold temperatures, and injection or transfer pressures, must be adjusted to minimize any significant bleeding, or continuous flow of material out the vents. With unusually soft flowing materials, such as some low viscosity epoxies, vents may be shallower than 0.003″, but should be proportionately wider, where possible and depending on the cavity volume, to insure rapid evacuation of the cavity.
7. The "manifold"—the space outside of the cavity vents, and inside the O-ring in the mold parting faces—will generally be large enough to prevent any flash going from cavity to the vacuum system.

8. Ejection pins, core pins, and inserts to be molded in, are sources of in-leakage of air when the cavity is evacuated. Pins should be "scored" with a groove completely around the circumference, at a point below (or above) the cavity, but close enough to the cavity that the "score" is exposed when the pin is in the full "up" position. The score should have a diameter about three times the nominal clearance between the pins and its hole, and with a half-round cross section. In operation, the "score" will generally fill with material and serve as an O-ring seal after the first shot. No effort should be made to "blow off" this material in the score between cycles.
9. If the cavity is to accept inserts, which are to be molded in the final part, provision for vacuum venting outside of the inserts should be considered, not only to prevent air in-leakage should an insert not fit perfectly, but also as further insurance of good plastic flow around the insert(s) and good bonding between the plastic and insert, free of any weakening voids. Extra vacuum paths from a sealed space behind the inserts, leading to the manifold between mold halves or to the vacuum port on the outside surface of the mold, can generally be included in the original mold design and manufacture.
10. It should be noted that specific values given above are guidelines and may be varied in specific instances.

With All the Above Provisions, How Good is the Vacuum in the Cavity?

Based on experience with over 300 production molds that Hull Corporation has supplied to the plastics industry, including both injection and transfer molds, we believe that the cavity vacuum is no less than 20″ Hg., at the instant material enters the gate. This pressure is maintained, or even reduced—because of the above-described design provisions—until the last vent is "sealed off" by material in the cavity. After that time, cavity fill is completed with such remaining "rarified" air being compressed by the inflowing plastic. The mass of such trapped air is generally small and insufficient to prevent complete cavity fill. While it is generally not possible to measure air pressure inside a cavity during production conditions, we feel that vacuum in the reservoir should not fall below 25″ Hg., even during the initial inrush of air from the cavity at the moment of vacuum solenoid valve actuation. Under those conditions, and with the remainder of the vacuum provisions as described above, voids from entrapped air should be non-existent.

If voids do develop, despite all the above conditions being fully met, we generally find that some parts of the mold or inserts are "outgassing"—residual solder flux, an outgassing adhesive, air entrapped in the fine interstices of a wire winding, excess mold lubricant, machining oils not cleaned from surfaces of mold cavity inserts, etc. All these "oversights" have actually occured in our experiences over three decades of furnishing vacuum vented mold systems.

6.9 TRANSFER/INJECTION CAVITY LAYOUTS

The ultimate goal, as earlier mentioned, is to achieve maximum molded density (MMD), and one of the factors to be considered in transfer or injection molding is the need to get the hot compound into the cavities as quickly as possible to avoid any precuring of the compound as it enters the hot mold and passes through the runner and gate system. The shear effect on the compound will tend to elevate the temperature quite rapidly, especially as it goes through the much smaller gate area, radically accelerating the curing reaction.

An effective deterrent to overheating the compound is to design the mold with runners that are as short as possible and of equal length and configuration. The layouts described below are some of the ways this can be accomplished. Runner and gate sizes will depend a great deal on the compound in use, and the examples in Figure 23 can serve as guidelines for the design of runners and gates for the various compounds.

Runner and gate configurations depend a great deal on the compound type being molded, the cavity layout, and molded part configuration. General purpose compounds, usually containing fillers and/or reinforcements that are small in particle size, do not require large runners or gates, whereas the high-bulk factor, high-strength compounds with bulkier reinforcements need runners and gates of sufficient size that they do not severely degrade the physical properties of the molded part. The runners and gates must be of a size that will allow the reinforcements to easily enter the cavities and will also make it possible to fill the cavities in a reasonable time period.

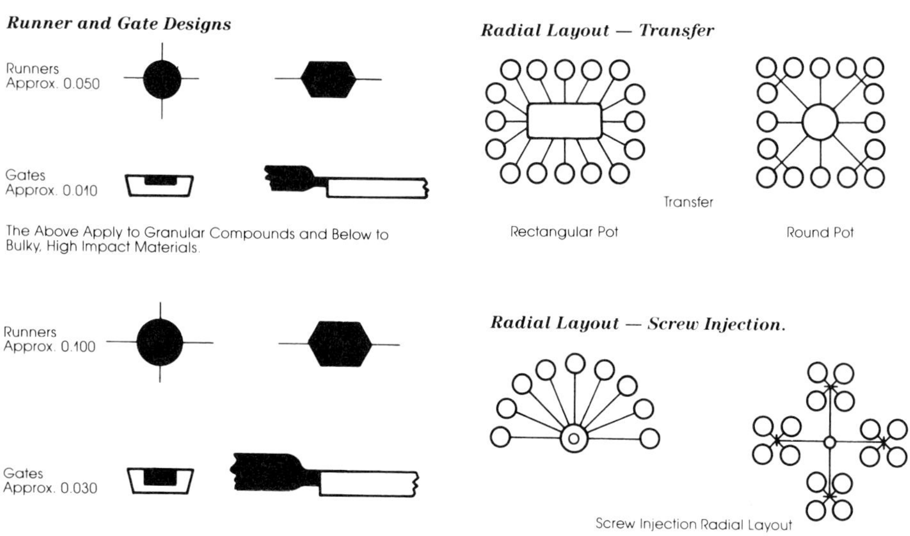

Figure 23. Layouts for molds with runners.

The basic reason for using high-strength reinforcements is to provide the needed reinforcement and the use of small area gates will result only in strong runners and weak parts. All molded parts using high-strength reinforcements will require a gate removal operation, usually done at the press during cure time. There is little, if any, added cost in removing a 1/8-in.-thick gate as compared with a gate of 1/16 in. thick, whereas the 1/8-in. gate will better achieve MMD.

6.10 MOLD STEELS

The following is a list of high-grade tool steels recommended for use in the various types of molds:

Machined Forces and Cavities:

Low carbon steel. Can be case hardened 3/32 in. deep and easily machined.
Oil hardening tool steels. Hard surface, tough strong core, good machining.

Mold Plates—retainers, clamping, and support plates:

Hot rolled steel. Easy to machine and may be carburized with low distortion.

Prototype Tooling:

Water hardening tool steel. Good machinability, low cost.

Hobbed Cavities:

Pure iron carburizing steel, very soft center allowing ease of hobbing. Can be water hardened with a surface hardness 3/32 in. deep.

Hobs:

Air hardening tool steel. Very hard and strong when heat treated. Excellent wear resistance.
Stainless steels are not recommended for compression molds but will perform well in transfer and injection molds. Limited in surface hardness but hold a good polish.

Note: Hobbing is a technique for producing duplicate cavities by making a hardened steel master (duplicate of the molded part) that can be forced under very heavy pressure into a soft steel blank, forming a cavity blank, which then is further machined to fit into the mold frame. Hobbing is a very inexpensive method of producing duplicate cavities of relatively simple configurations.

Table 1. Tooling and design information when constructing molds for use with silicone molding compounds.[5]

Tooling considerations[a]	Compound Number 306	307	308	480	1-5021	Test method
Steel	←		Tool steel		→	— —
Hardening	←		55–60 Rockwell C		→	— —
Plating	←		Chrome plating		→	— —
Gate depth (mils)	12–18	12–18	12–18	12–18	12–18	— —
Vent depth (mils)	2–3	2–3	2–3	2–3	2–3	— —
Runner depth (in.)	1/8	1/8	1/8	1/8	1/8	— —
Runner width (in.)	3/16–3/8	3/16–3/8	3/16–3/8	3/16–3/8	3/16–3/8	— —
Runner length (in.)	8–10	6–8	8–10	6–8	6–8	— —

[a]Information provided only as guide for building tooling.

REFERENCES

1. Hull Corp., Hatboro, PA.
2. New Dynamics Corp., Ivyland, PA.
3. Dusan Equipment Corp., Gilroy, CA.
4. Hull Corp., Hatboro, PA.
5. Dow Corning Corp., Midland, MI.

CHAPTER 7

COST ESTIMATING

7.1 MOLDED PARTS

Estimating the costs involved in the production of thermosetting molded parts begins with a "Request for Quotation" (RFQ) from an internal sales/marketing group or from a similar group in an outside customer's organization. The RFQ will contain the following:

- Detailed part drawing specifying

  Compound
  Inserts
  Dimensional requirements
  Cosmetic requirements

- Production quantity (total)
- Production schedule (monthly/quarterly/annually)

With the above information, the estimator can calculate the labor and material costs based on the following:

Labor

Molding method
Number of cavities
Postmolding operations
  Flash removal
  Tumbling
  Buffing
  Polishing
  Painting
  Plating

Inspection
Packaging/packing/shipping

Material

Part and shot weight
Compound cost
Insert cost
Decorative materials
Packaging

Note: For part weight, see the chart on pages 185–186.

Scrap: The scrap cost will depend on how much labor and material has been expended on the scrapped part. A scrapped part after molding will cost less than one scrapped after molding and finishing operations have been completed. In the event that inserts have been used, the cost of insert recovery should be balanced against the initial insert cost. Precious metal plated inserts are often worth recovery.

There is great need for the estimator to be in touch with all manufacturing personnel responsible for the various operations involved to secure their expertise and experience before the quotation has been finalized.

The application of factory, marketing, and managerial overhead or burden costs as well as profit margins will depend on individual accounting practices. There are a number of computerized systems for accurate and rapid calculations of manufacturing costs associated with the molding of thermosets that make it possible to quickly and easily arrive at optimum conditions when estimating molded part costs.

7.2 MOLDS

Mold cost estimating can be done "in house" if the individual doing the estimating has had considerable experience with the design and construction techniques involved in mold making. Very often, molders will depend on their mold makers on the outside to supply them with firm quotes based on an assembly drawing of the mold or a fairly detailed sketch describing the important features of the mold design. Regardless of who actually does the cost estimating—in house or by the outside mold shop—there are basic factors that are the core of the estimate:

- Design hours
- Raw materials
- Construction hours
- Assembly hours

The above factors will vary according to the following:

- Type mold

  Compression
    Flash type
    Positive type
    Semipositive
  Transfer
    Pot type
    Plunger type
  Screw transfer
  Screw injection

- Number of cavities
- Design complexity

  Cam action side draws
  Hydraulic/air action side draws
  Stripper plate action
  Runner/gate design
  K.O. system
    Standard pin plate
    Hydraulic/air actuated

- Cavity fabrication

  Hobbing
  Machining
  Duplicating
  Electric-discharge machining (EDM)

- Force fabrication

  Machining
  Duplicating
  EDM

7.2.1 Design Hours

Mold design costs are generally figured on an hourly rate with these rates varying from $25 to $45/hr depending on the design technique employed. Manual designs off a drawing board are at the lowest hourly rate, whereas design employing computer-aided equipment (CAD) will carry a higher rate. The CAD system will provide valuable extras such as tool path dimensions for numerical control machines that are derived from the original molded part design with the appropriate shrinkage figures calculated into the tool paths.

Mold design costs are often considered to be approximately 10–15% of the overall mold cost, but this will depend on where and how the mold design is performed.

7.2.2 Raw Materials

The raw material costs for mold components such as the cavities, forces, runner, and gate blocks will be minor compared with the fabrication costs involved in transforming the steel into a working mold component.

The primary raw material cost will be the mold frame itself, and this cost is easily determined from the catalog of any reputable mold base supplier. The mold base will consist of

- Retainer set
- Backing plates
- Clamping plates
- K.O. and return pins
- K.O. assembly
- Pot/sprue bushing
- Heaters
- Cam locks
- Rack and pinion sets

In all, the raw material mold costs will amount to approximately 10–12% of the overall mold cost.

7.2.3 Construction Hours

Here is where the real costs come into focus because here is where the man-hours start to accumulate. Also, here is where the quality of the mold is determined. The workmanship and accuracy built into the mold will govern its performance over its working life.

Some of the more important pieces of mold making equipment that will be utilized in the construction of a mold are

- Lathes
- Milling machines
- Surface and cylindrical grinders
- Radial drills
- Jig borers

- Shapers
- Electrical-discharge (EDM)

Since the forces and cavities are the key components producing the final molded part, their method of fabrication has to be carefully considered.

If the mold is to contain several of the same cavity and the part design is simple—lending itself to the "hobbing" process—the fabrication can be relatively inexpensive. Hobbing is a process of forming a mold cavity by forcing a hardened, highly polished steel master of the identical shape and dimensions of the outer periphery of the desired molded part into a soft steel cavity blank. The master will have been machined with the appropriate shrinkage dimensions. The hob is capable of producing a large number of cavity blanks—all identical in their cavity dimensions and ready for shaping, grinding, and hardening prior to fitting into the mold frame. Hobbed cavities can be used in all thermoset molds providing the cavity design is suitable.

It is also possible to fabricate the cavity and force components by the duplication process—again depending on the suitability of the part design to this process. Duplication can be done with a preformed master or duplicate of the desired part. The master is set up in a vertical milling machine utilizing a sensitive table and a tracing spindle head, which combine to permit the reproduction of nearly any three-dimensional or flat profile. Again, the master is capable of producing a number of duplicate cavities or forces at a relatively low cost.

Duplication is also possible with simple configurations when using the electric-discharge technique (EDM). The master electrode made of carbon or copper is produced in the appropriate shape and placed over the cavity block in an EDM machine. The electrode is slowly lowered into the cavity block with the electrical discharge eroding the surface of the block in the shape of the electrode. The usual method is to use three electrodes—rough for the initial cut, smooth for the second cut, and fine for the final cut. The chief advantage to the EDM technique is that it can be done with hardened cavity blanks, eliminating the hardening process. The fine surface attained after the third cutting still has to be polished to produce the proper surface for hard chrome plating.

More complicated and complex cavity/force configurations will require that the machining be done with milling cutters on a one at a time basis, with a systematic procedure that will complete a specific tool path on each cavity/force before proceeding to the next path.

As earlier mentioned, mold costs will vary according to the type of mold being contemplated. For example, compression molds will vary in cost depending on the cavity design. Simple flash type molds are the least costly, whereas the positive and semipositive prove to be the most costly due to the need for close tolerance construction in the telescoping force and cavity areas. Transfer and screw injection molds will normally cost more than compression molds due to the need for sprue bushings/pots/plungers, runner blocks, gate inserts, etc. Also the ability to incorporate cam action side draws, while reducing molded part costs, will add to the cost of closed molds such as transfer or injection types.

7.2.4 Assembly Costs

Final assembly of the mold components into the mold frame is preceded by highly polishing all molding components and sliding units along with the final grinding and fitting of each component into the frame. This is a very slow, tedious, and costly but extremely important phase of mold making.

Thermoset molds, whether compression, transfer or injection type, are constructed with high grade tool steels with all sliding components, forces, cavities, and runner blocks hardened, polished, and hard chrome plated.

The mold is one of the key elements in the entire process of molding thermosetting molding compounds and it is extremely important that great caution be exercised in the design and construction of the mold. It is very poor economics to skimp on the quality of a mold because a poorly designed mold with less than the best materials and workmanship will be a constant expense, far outweighing any cost saving anticipated.

7.3 FINISHING TOOLS

Cost estimating for such tools and fixtures as are needed for clamping or cooling devices when the parts are ejected from the mold follow the same pattern as for mold making. The level of skill is not as high, but there is a need for accuracy and good workmanship. Time and materials will determine cost, generally on a machine shop hourly rate—somewhat lower than for mold making.

Very often finishing fixtures used for gate removal or drilling operations are produced after the mold has been approved and shots from the mold serve as guides or patterns for the toolmaker during construction.

It is important for the cost estimator to be sure that finishing tools and operations are thoroughly covered prior to finalizing the cost estimate.

CHAPTER 8

PRODUCT DESIGN

The family of thermosetting molding compounds offers the product designer a wide range of compounds from which to choose for any given application. With the rapidly changing technologies currently demanding higher performance at lower costs, an ever greater emphasis on accuracy in the material selection process becomes paramount.

8.1 PRODUCT DESIGN GUIDELINES

The process of molding thermosetting compounds into desired configurations places certain basic design considerations on the product designer. Some of the more important factors to be carefully analyzed are

- Tapers and surfaces finishes
- Wall thicknesses
- Use of bosses/barriers/ribs/shoulders
- Parting line location
- Ejector pin location gate location
- Molded-in side holes/undercuts
- Fiber orientation

8.1.1 Tapers and Surface Finishes

All molded thermoset components must have sufficient taper on those molded surfaces parallel to the mold opening direction—generally 1/2–3/4° per side, depending on the compound being used, with all dimensional tolerances taking the taper into consideration (Figure 1). Lack of sufficient tapers or draft will result in parts that are very difficult to eject from the mold, causing warping and cracking and necessitating longer molding cycle times. The same taper considerations are applicable to any and all side draws or core pullers.

Surface finish also contributes to part ejection, as well as part appearance. The mold surface produces the part finish, so the designer must specify the

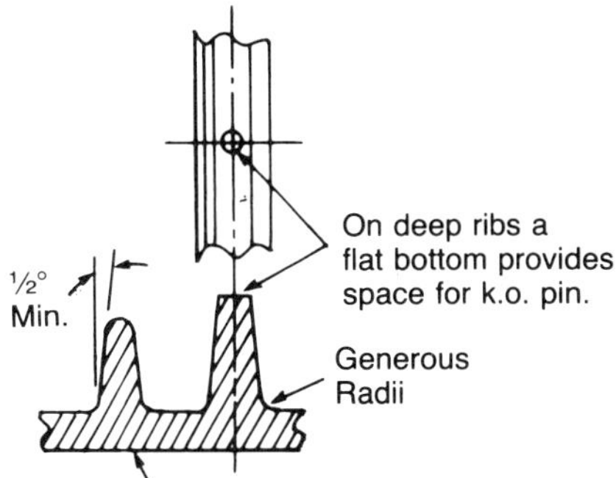

Figure 1. Tapers/mold draft.

type surface desired if it is of major concern. Special surface effects such as a matte finish for the molded part will have to be indicated by the mold designer.

8.1.2 Wall Thickness

Regardless of the compound or molding process, cross-sectional wall thickness should be kept uniform and to a minimum, consistent with desired property requirements (Figure 2). Wall thickness determines cure time, which, in turn, relates directly to overall cycle time—productivity and cost. The thicker the wall, the longer the cure, sometimes necessitating a postbaking operation. Careful thought must be given to this phase of the product design because cure time, along with molded density, greatly affects the ultimate mechanical, electrical, chemical, and thermal properties of the molded article.

8.1.3 Bosses/Barriers/Ribs/Shoulders

All bosses, barriers, ribs, and shoulders must be designed with molding constraints in mind (Figures 2 and 3):

- Uniform wall thicknesses consistent with product design needs.
- Generous use of radii—no sharp corners. See examples below.

8.1.4 Parting Line Location

Location of the parting line on the periphery of a compression molded part is of greater importance than the parting line on a transfer or injection molded

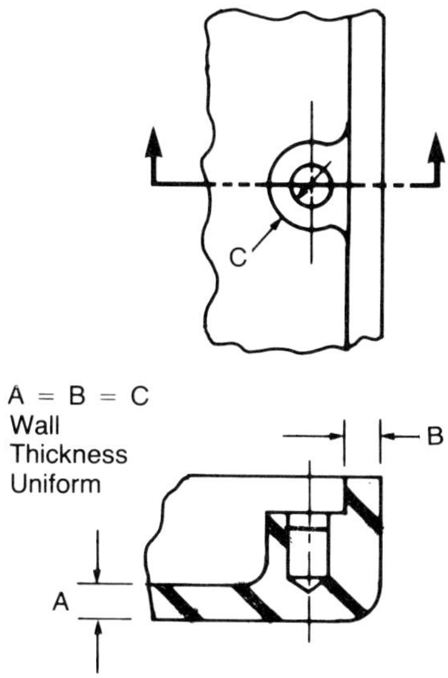

Figure 2. Cross-section of wall thickness.

part (Figure 4). There will always be flash, requiring removal, on the parting line of a compression molded part, whereas a molded part from a well-designed and constructed injection or transfer mold will generally be relatively flash free, particularly if the mold is well maintained. Flash on an irregular parting line on a compression molded part will be more difficult and costly to remove, especially if the compound used is a high-strength, high-bulk factor product that will produce parts with parting line flash thicknesses in excess of 0.005 in.—very difficult to remove. All compression molds will have flash in through molded holes with the flash thickness varying from 0.005 to 0.015 in. depending on the compound used. The usual tumbling or wheelabrating flash removal systems

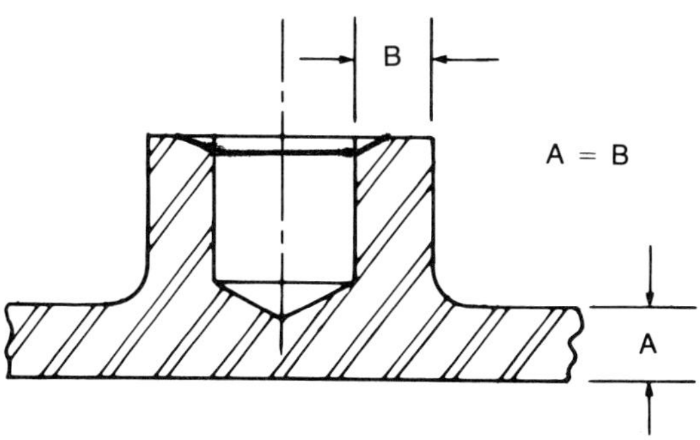

Figure 3. Stand off boss.

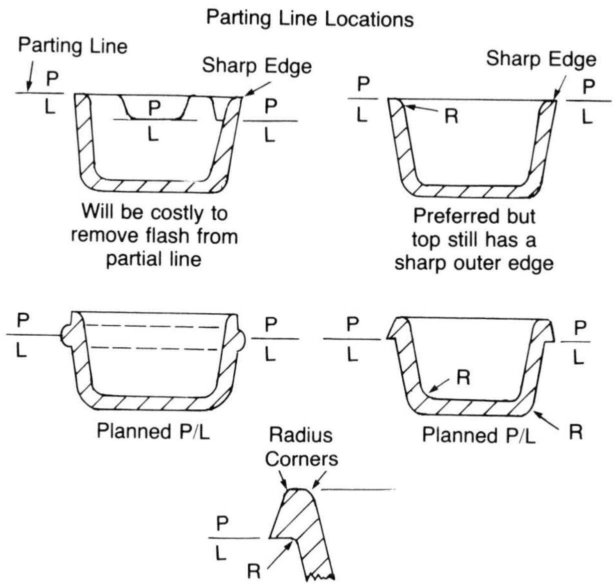

Figure 4. Parting line locations.

are effective on general purpose compounds (flash thickness 0.005 in. or less) but ineffective with compounds producing flash of greater thickness.

8.1.5 Ejector Pin Location

Ejection of a molded part from the mold is usually accomplished with pins strategically located within the mold cavity or force—sometimes with K.O. pins located on both halves of the mold. These pins are mostly round, but occasionally will be square or even rectangular in shape. It is not unusual for an entire mold section to be used as an ejector.

Uniform thrust on the molded article during ejection is most important to avoid any stress or strain that might cause warpage or out-of-flatness. As mentioned earlier, the mold surface condition, proper mold draft or taper, degree of cure, and amount of mold lubricant or release agent will have a marked effect on the ejection phase of the molding operation. Again, it is important that the mold surfaces be highly polished in the direction of the mold opening to ensure proper release.

In all multicavity molds, the individual cavities should be identified by number or letter, with the identification put on the molding end of an ejector pin located in an easily seen area of the molded article. Cavity identification is a necessity in any quality control system.

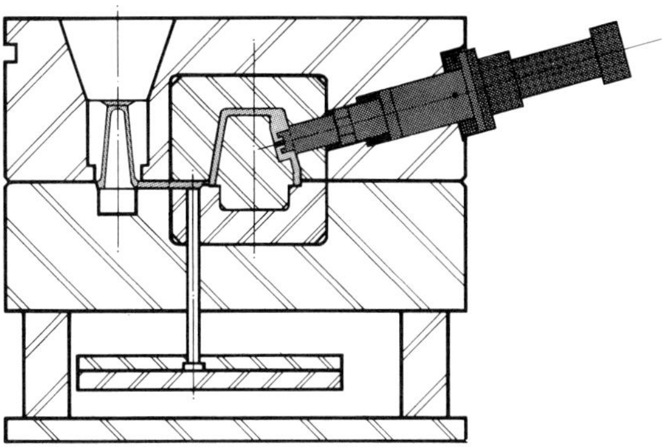

Figure 5. An independent locking device is usually hydraulically or pneumatically operated.[1] See page 173.

8.1.6 Side Holes/Undercuts

8.1.6.1 Compression Molds

The molding in of holes or molded sections that are not parallel with the mold opening is difficult with compression molding because any moving mold components will be exposed to the compound flow as the mold is closing and will not be in their final location until the mold is completely closed. This can be done with the use of independent locking devices—usually hydraulically or pneumatically operated as shown in Figure 5, which will seal off the moving components. Such devices are costly and do not always provide flash-free parts.

8.1.6.2 Transfer/Injection Molds

The closed mold design exemplified by either the transfer or the injection method provides a far simpler and safer means for molding side holes and/or undercuts. The mold is fully closed with all of the side draws locked in place prior to any compound entering the mold. The molded holes, parallel to the

A quantitative prediction of fiber orientation distribution in molded parts is not yet possible. However, several empirical rules have been established which can be used for qualitative predictions of the likely orientation. These rules, combined with a simple analysis of the mold-filling pattern, can provide useful guidance to the proper fill rates, gating, venting, and other changes in part geometry needed for optimal part performance.

Compression of the material causes the fibers to align perpendicular to the direction of mold closing even when there is little lateral flow. In this case, random orientation in the plane perpendicular to the mold force is favored except when the fibers are constrained by the mold geometry.

Material flow along the mold produces rapid changes in orientation. Converging flow tends to produce alignment in the bulk flow direction, while diverging flow results in transverse orientation.

In straight channel flow, if the flow path is sufficiently long, surface layers with fiber alignment parallel to the flow direction are likely to form. Surface layers are possible even when the flow geometry is strongly divergent. In the center regions of the moldings, fibers are generally predominantly oriented transverse to the flow. The orientation will also vary with the section thickness. As thickness decreases, the fibers in the transversely oriented core become more highly aligned in the plane of the molding; for thin sections, the core disappears altogether.

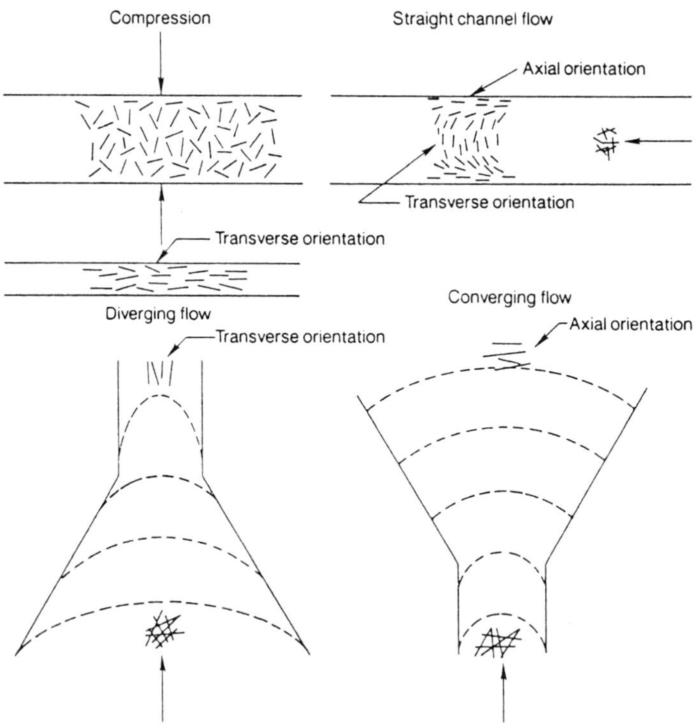

Figure 6. Rules of thumb for predicting fiber orientation distribution.

parting line or not, can be molded flash free; in addition, any inserts within the mold can be properly sealed off and molded essentially flash free.

8.1.7 Fiber Orientation

All thermosetting molding compounds contain fillers or fibrous reinforcements—often as high in content as 50–75% of the total compound weight. These materials range from wood flour, talc, and cotton fibers to glass and organic fibers of various lengths.

During the molding process, these reinforcements will tend to "orient" themselves in various ways—largely dependent on

- The type and size of the reinforcement
- The molded part configuration as it relates to the flow pattern of the compound within the confines of the mold
- The rheology of the compound
- The molding process—compression/transfer/injection
- The gate location and configuration in the case of either transfer or injection molding

The main role of reinforcements in a compound is to impart strength—primarily impact and flexural—to the molded article. The ultimate properties of fibrous reinforcements are directional so that their orientation within the molded article during molding will greatly affect the article's final physical property profile. If the fibers are unidirectional, tensile and modulus will be maximized, while a transversal alignment of the fibers tends to reduce those values. On the other hand, a random orientation will result in property values uniform in all directions but not as high as is achieved with a unidirectional orientation. Fiber orientation can also affect molded dimensions because the alignment of the fibers within a molded part will influence thermal expansion, a principal cause of molding shrinkage. It is primarily the resin that shrinks, but its direction and amount of shrinkage are highly dependent on the reinforcement and its final orientation within the molded part (see Figure 6).

The flow of the compound during molding—carrying the fibers with it—is greatly predetermined by mold design features such as

- Molded part configuration
- Runner and gate configuration with closed molds
- Gate location

Once a clear understanding of the *total* requirements has been established, the search for the best available molding compound will come into better focus, freeing the product designer to investigate some of the other factors that will influence the ultimate compound selection even further. Some of these factors are

- Molding Process
- Use of inserts
- Performance testing
- Prototyping

8.2 MOLDING PROCESS

It is not necessarily the product designer's province to select the exact molding process, but it is important that the designer be familiar with the different processes available and how each process affects part design and part performance. It is usually in the designer's interest, early on, to be in contact with the *compound suppliers, the molders, and the mold designers* to enlist their respective expertise to confirm the choice of compound and to determine the proper molding procedure along with other mold design considerations such as

- Parting line location
- K.O. pin locations
- Number and location of cavities
- Gate Location
- Draft
- Radii
- Wall thicknesses
- Runner location and configuration

The best time to settle these important questions is during the design stage, not *after* the mold construction has begun or, worse yet, after the mold construction has been completed. Mold revisions are very costly and time delaying. A few days or weeks spent clarifying design criteria will stave off months of very costly lost time correcting design errors after the mold has been completed and is ready for sampling.

8.3 INSERTS

8.3.1 Molded-in vs. Postmolding Insertion

The use of inserts, generally metal, in the design of thermoset molded parts is common practice. They provide the product designer with very practical methods for assembly purposes and for the installation of electrical contacts in the molded part.

Thermoset moldings are readily machined, drilled, and tapped with standard equipment and tools. However, the employment of metal inserts with male or female threads, inserts with solid, round, or flat contact surfaces, or tubular inserts of various configurations will often serve the purpose more effectively and at less cost—particularly when they eliminate postmolding operations. In

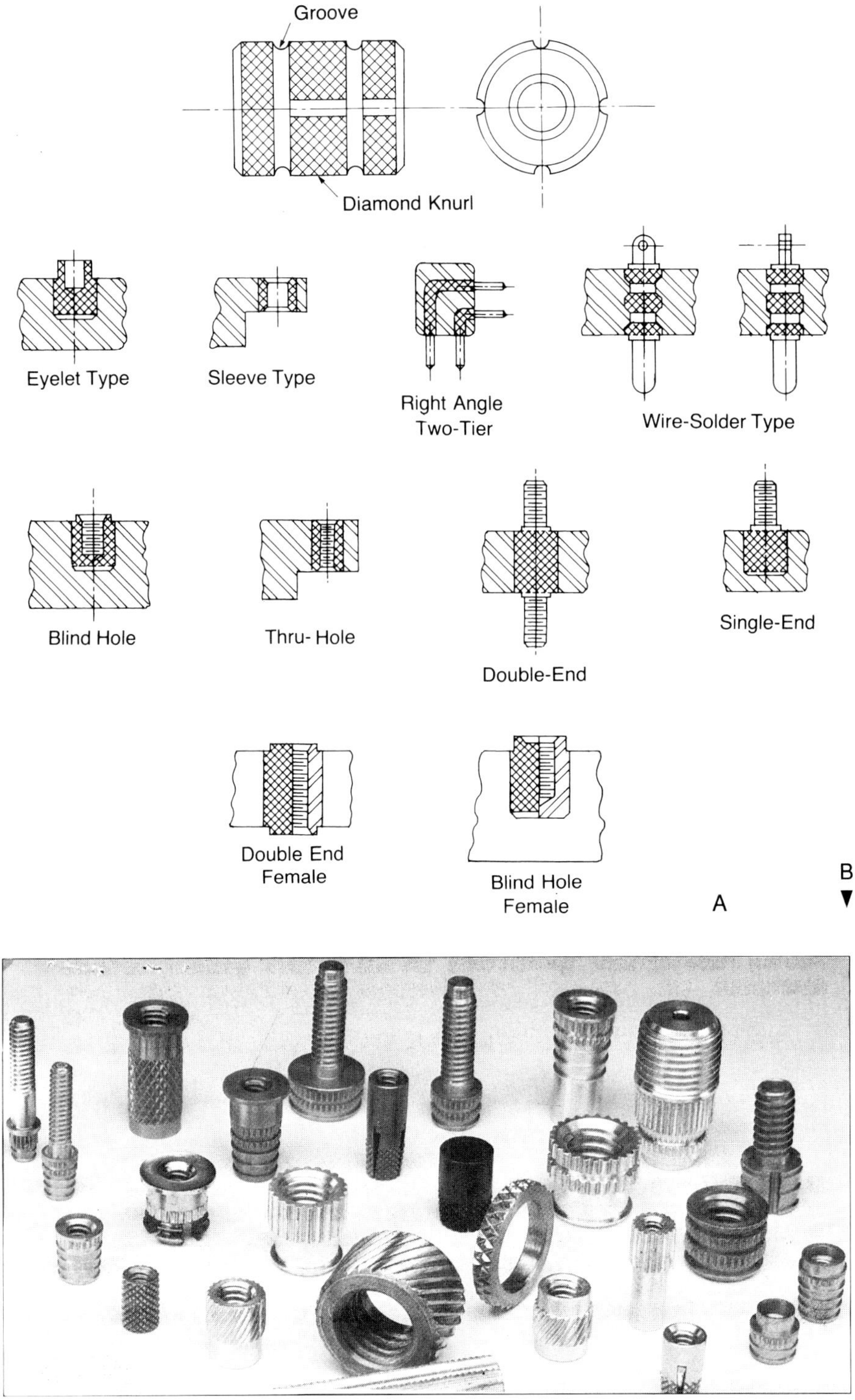

Figure 7. Types of molded-in inserts.

designs that require frequent assembly and disassembly, the use of molded holes and self-tapped screws should be avoided in favor of either molded-in or postmolding inserted devices (Figure 7A and B).

The generally accepted anchorage techniques employ the use of end-knurls, ranging from fine to coarse and usually of the diamond pattern. Lengthwise grooves help to prevent rotation of the insertion while grooves around the circumference prevent pullout. Care must be taken to avoid having sharp corners or edges in the grooves.

8.3.2 Insert Dimensions and Tolerances

The vast majority of mold-in inserts are produced on automatic screw machines and are made of various materials such as copper, brass, bronze, aluminum, and stainless steel.

Cold forged inserts produced with rolled threads have been used where tolerances levels are more open. Materials suitable for cold forged inserts are brass, copper and copper alloys, stainless steel, and coin silver. This type insert can be produced to tight length and diameter tolerances but at the cost of added machining operations.

Many of the above insert styles or types can be plated for special applications, generally to meet electrical conductivity and/or corrosion resistance requirements.

The decision to use or not to use molded-in inserts will often depend on

- Ultimate number of parts to be molded
- Quantity and variety of inserts in each part
- Number of cavities in the mold

The total number of inserts to be loaded into a mold during the "open mold" segment of the molding cycle will obviously have a direct bearing on the total cycle time and, consequently, on the molding cost. There are effective ways to reduce the "open time" at the press, but they will bring with them some added costs in special equipment and/or increased mold and tool costs.

8.3.3 Insert Loading Devices

Such devices range from simple fixtures that will hold several inserts in their proper alignment so they can be manually positioned over the cavity and allowed to drop into place in a single operation to very complex automatic insert loaders involving rotary feeders that pre-position the inserts and then feed them into either a manual loader or directly into the lower half of the mold itself.

The simple, manual insert loading device can be loaded by the operator during the curing cycle for the next shot and will be the least costly method

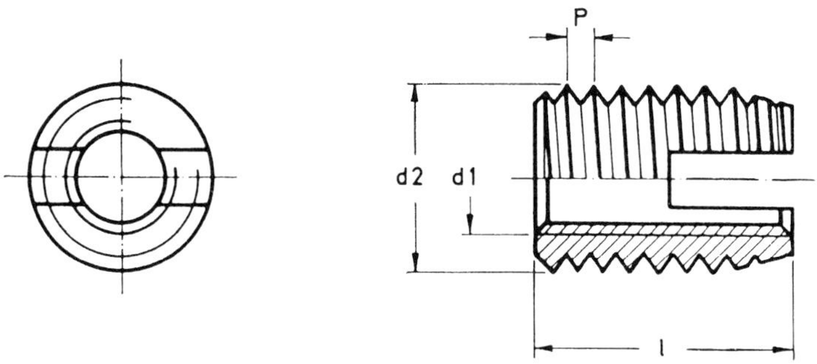

Figure 8. The Ensat 302 threaded insert with cutting slot is a self-cutting connecting device for producing wear and vibration resistant screw connections which can be subjected to high loadings in materials which have a low shear resistance, such as light metal alloys, cast-iron, brass, bronze, NF metals, plastics, laminated materials, and hardwoods.

providing there are a limited number of inserts to be loaded or a limited number of moldings to be produced.

Another approach to insert molding involves the use of a mold designed with one upper mold half and two identical lower halves. Such a mold has been detailed in Chapter 3, Section 3.4 (Shuttle Press Molding).

These molding techniques are commonly used when the insert loading time will greatly exceed the curing time due to the number or complexity of the inserts being loaded.

The decision to utilize any of these "open time" reduction methods has to be considered in light of the economics of each given situation. In any insert molding operation, there is an "optimum" number of cavities with production volume being the deciding factor. Obviously, a molding situation calling for monthly production of 100,000 parts over a 1- or 2-year period will allow the use of more extensive mold and tooling expenditures than a molding situation calling for a monthly production schedule of 10,000 parts over a 6-month term. The cost of molds, tools, and fixtures can be amortized in the part cost more economically as the total volume of production increases.

8.3.4 Postmolding Inserts

The product designer has another option when considering the use of threaded inserts in a molded part—that of installing the inserts *after* molding. There are two basic insert types:

- Expansion type—generally internal threads that are inserted into a molded hole(s) when the molded part is cold. When the internal thread becomes activated with a male screw, the outer shell expands and becomes locked in place (Figure 8).[2] See page 173.

- Male or female threaded inserts with external spiral knurls that can be force-fitted into a molded hole(s) as soon as the part is removed from the hot mold. With this force-fitting method, the molded part is removed from the mold and immediately positioned into a fixture containing the inserts, which are aligned with their mating holes in the molded part. The inserts are pressed into the part, and, as the part cools, it shrinks onto the inserts, securely anchoring them in place (Figure 9).

Both methods are beneficial in that they eliminate the more costly molded-in insert techniques and are always free of flash on the threads. The expansion type can be inserted in a secondary operation, not affecting the molding cycle, whereas the force fitting method can be utilized during the molding cure time.

Once again, the decision to utilize either the postmolding insertion or the molded-in inserts—as opposed to drilling and tapping—will depend a great deal on the proposed volume of production. It is wise for the designer to discuss these types of decisions—as they affect part design—with molders, mold designers, and compound suppliers before freezing the design.

8.4 PERFORMANCE TESTING

Which performance property or combination of properties will most accurately predict the ability of a design to meet its end use requirements?

There are several standard tests commonly employed to measure

- Creep
  - Strain
  - Modulus
- Rupture
- Insulation resistance
- Volume/surface resistivity, ASTM D-257
- Dissipation factor, ASTM D-150
- Dielectric strength, ASTM D-149
- Dielectric constant, ASTM D-159
- Arc resistance, ASTM D-495
- Flame resistance, U.L. 94
- Coefficient of thermal expansion, ASTM D-696
- Deflection temperature, ASTM C-648A
- Glass transition temperature (T_g)[1]

---

[1] T_g is not a specific temperature but is expressed as a temperature range. From the onset through the upper end of the transition, the temperature rises by 20–50°C. T_g could be described as "the limit of a molded specimen's ability to bear a continuous load and remain dimensionally stable."

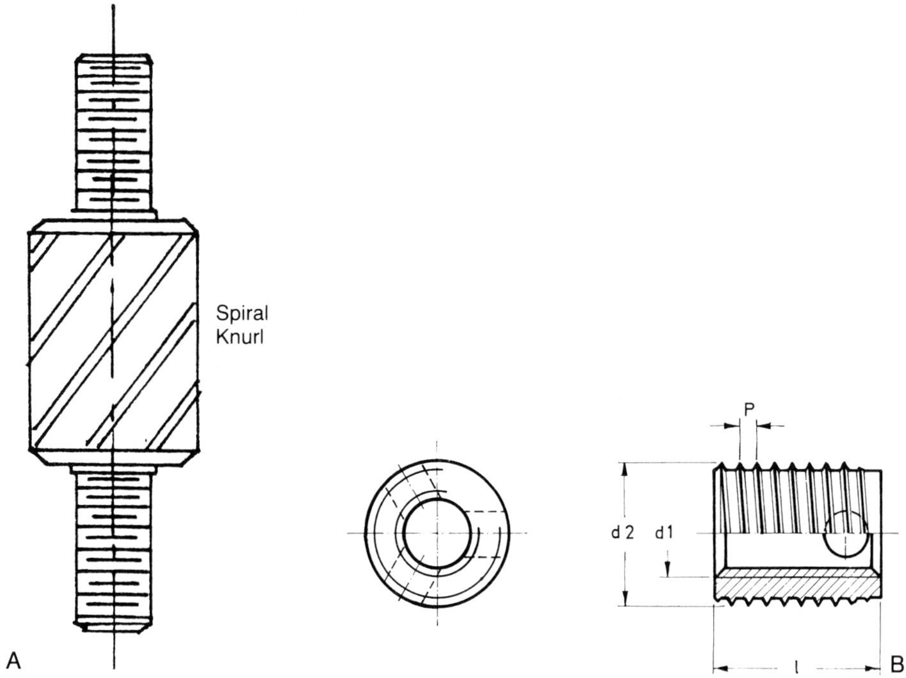

Figure 9. The Ensat threaded insert with cutting bores is a self-cutting connecting device for producing wear and vibration resistant screw connections which can be subjected to high loadings in materials which have a low shear resistance, such as aluminium and aluminium alloys, magnesium alloys, duroplastics, and thermoplastics (with the exception of rubber-soft thermoplastics, Shore A hardness < 100).

It is important for a design engineer to recognize this phenomenon, particularly as it relates to the elevated temperature performance of the product. Thermosets, due to their chemical cross-linkage resulting from the chemical reaction during molding, retain significant mechanical properties *above* their T_g, resulting in excellent thermal stability.

In addition to these standard tests, the design engineer might also consider some performance tests that will simulate, as nearly as possible, the actual environmental conditions to which the molded part might be subjected. Here, ingenuity, coupled with expert engineering knowledge and experience, will be most helpful.

Sometimes a simple drop test of sufficient height onto a particular surface will determine the ability of the part to pass its specific quality specification. Or testing may require a very sophisticated procedure involving elevating temperature and humidity with increasing electrical and/or mechanical loads over extended time periods to determine the ability of the part to survive successfully in its intended end use.

Practical common sense, coupled with sound engineering practices, will, more often than not, result in meaningful testing procedures.

8.5 PROTOTYPING

A prototype of the intended molded part can often be valuable in determining part configuration, weight, and moldability. Prototypes can be machined from a block of molded plastic, preferably of the same compound chosen for the molded part.

However, should the part itself be of such configuration as to be very difficult or impossible to fabricate, or if the final compound selection has not been made, it would be well to build a single cavity prototype mold to evaluate various compounds.

When a multicavity mold is contemplated, it is not uncommon to first build a multicavity frame but with only one finished cavity. This cavity can then be utilized to determine material selection, moldability, part weight, molding parameters, runner configuration, gate size, and location. Once the single cavity has produced acceptable parts, the balance of the cavities can be completed while the single cavity continues to provide limited initial production. Parts from this single cavity can often be used to evaluate fixtures for postmolding operations, packaging, and the setting of quality assurance procedures.

8.6 COMPOUND SELECTION

Checklist for manufacturing and end use requirements:

Dimensional Requirements:

Key dimensions, special fixtures/special gages
Flatness/warpage, cooling fixtures
Concentricity
Hole alignments
Parting line
Assembly tolerances

Cosmetics:

Color stability
Surface finish
Buffing

Polishing
Painting
Stamping

Safety Considerations:

Toxicity
Flameproofing
Fragility

Cost Considerations:

Compound cost
Mold and tooling costs
Fixtures/gages/jigs
Molding costs
Preforming/preheating/insert loading
Operator and machine cost
Finishing—deflashing/gate and runner removal
Postmolding operations
Drilling
Contact insertion
Postbaking
Tapping
Printing/painting
Stamping
Machining
Staking

After completing such an analysis, the product designer will be in a much better position to select the molding compound that best fits the requirements of the application with a greater degree of certainty, so that a successful thermoset molding will be the end result.

8.7 COMPUTER AIDED DESIGN (CAD)

Over the past decade, the use of computers in the design of molded plastic articles has accelerated due to the lowered cost of computer hardware as well as the proliferation of software programs covering almost every phase of the molding industry. This checklist can assist in the analysis of what a product may encounter in its passage from an idea to its culmination as a successful thermoset molded component.

Mechanical Load:

	Time	Load	Intermittent	Continuous
Compression		psi	< >	< >
Flexural		psi	< >	< >
Tensile		psi	< >	< >
Torque		ft-lb	< >	< >

Electrical Load:

	Time	Load	Intermittent	Continuous
Insulation resistance			< >	< >
Arc resistance			< >	< >
Arc/track resistance			< >	< >
Volume/surface resistance			< >	< >

Temperature Conditions:

	Time	°C	°F	Intermittent	Continuous
CTE compatibility				< >	< >
Thermal resistance				< >	< >

	Ignition Time	Burn time	Toxicity effects
Flame resistance			

Chemical Resistance:

Type	Time	Temp.	Intermittent	Continuous
Aliphatic hydrocarbons			< >	< >
Aromatic hydrocarbons			< >	< >
Oils/fats/waxes			< >	< >
Alcohols—monohydric			< >	< >
Alcohols—polyhydric			< >	< >
Phenols			< >	< >
Ketones			< >	< >
Esters			< >	< >
Ethers			< >	< >
Inorganic acids concentration/dilution			< >	< >
Bases concentration/ dilution			< >	< >
Salts—acid/neutral/basic			< >	< >
Organic acids concentration/dilution			< >	< >
Oxidizing acids concentration/dilution			< >	< >
Sunlight/weathering			< >	< >
Fully halogenated hydrocarbons			< >	< >
Partly halogenated hydrocarbons			< >	< >

The product designer has access to programs detailing comparative property values of practically every known thermosetting compound as a guide in compound selection, computer programs involving finite element analysis, and, to some degree, mold flow analysis. Mold flow analysis of thermost molding is still in the formative stages, with much more developmental work needed.

Such programs, prior to finalizing part or mold designs, can provide invaluable insight into

- Orientation of fibers during molding
- Proper fiber size and length
- Runner configuration
- Gate configuration and location
- Location of heating elements for optimum mold heating
- Size and location of vents to avoid knit lines
- Size and location of bleed-outs on compression molds
- Configuration and location of K.O. pins

Computerized modeling of mold filling, whether compression, transfer, or injection, has been and will continue to be a tool for increasing the understanding of what is actually taking place within the mold during the filling and curing stages of thermoset molding process.

The part design details developed and finalized on the part drawing are retained within the computer memory, available to the mold designer for his use and also to the mold maker as well as the tool and fixture maker for their use. The currently available CAD systems such as AutoCad and VersaCad provide invaluable assistance in the part, mold, and fixture designs.

Standard mold base and component manufacturers have software programs of varying detail and cost that enable the mold designer to utilize standard mold components already drawn to scale and available from the software library. Mold designing speed, accuracy, and detail have been greatly enhanced, especially when combined with molded part design data stored in computer memory.

References

1. Stilson Div. Stocker & Yale, Inc., Frasek, Michigan 48026.
2. Kerb-Konus-Vertriebs-GmbH, Postfach 1663, D-8450, Amberg, Germany.

RECOMMENDED READING

Bruce B. Fitts, *Fiber Orientation of Glass Fiber-Reinforced Phenolics*. Rogers Corp. MMD, Manchester CT 06040.

Joel Frados, "Design Standards for Inserts," *(SPI) Plastics Engineering Handbook*, 4th ed.

CHAPTER 9

FUTURE FOR THERMOSETS

9.1 NEW COMPOUNDS

The constant market demand for improved compounds to meet new and challenging applications will continue to bring forth innovative insights into every facet of the thermosetting compounding industry. The formulations, processing techniques, production equipment, and quality control will all benefit from the following:

- Increased technical understanding of the art of compounding
- The use of computer technology in compound formulations
- Recognition and correction of product deficiencies
- Implementation of statistical process control techniques to maintain high-quality standards
- Increased use of computerized process controllers on production equipment to improve process speed and consistency.

New formulations will be taking more advantage of the benefits of alloying/blending with other polymers to gain greater "toughness" without sacrificing the superior thermal properties already present in thermosets. This, coupled with an increased awareness of new, improved reinforcements, will form the base for continued growth well into the next century.

9.2 NEW PROCESSES AND EQUIPMENT

The current high production, automated compounders are proving to be forerunners of future equipment designs. They recognize the need for speedier processing in line with increased consistency and will be more equipped with on-line computerized controllers to achieve those goals. With greater emphasis on the use of various reinforcements to achieve higher physical strength values, the need for production equipment or processes that will preserve the integrity of the reinforcement system becomes paramount.

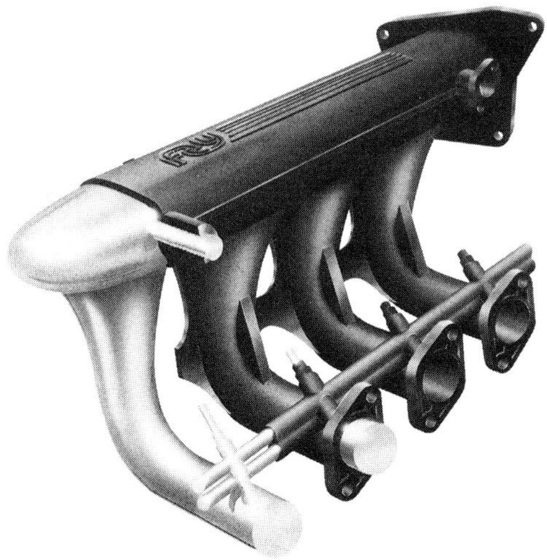

Figure 1. Automotive manifold housing molded with melted core technique.

9.3 NEW MOLDING TECHNIQUES AND EQUIPMENT

Resin transfer molding (RTM) and structural reaction injection molding (SRIM) are invaluable additions to the thermosetting molding industry, providing molding methods that are most suitable for highly reinforced compounds. Both systems are capable of automation for large-scale productivity and are also capable of producing very large area moldings. Again, the key to expansion of automated production facilities lies in the volume of moldings required.

Another important growth area for the use of thermosetting molding compounds lies in the metal replacement field—particularly where weight reduction with little or no loss of strength is of primary consideration. The use of meltable cores to mold hollow, one-piece automotive engine components such as water pump and manifold housings with either transfer or injection molding techniques will bring about renewed interest in this and other similar application areas (Figures 1 and 2).[1]

9.4 NEW APPLICATIONS

There will continue to be a greater recognition by design engineers in the automotive, aircraft, aerospace, construction, transportation, and marine industries of the benefits to be derived from the specifying of thermosets in those applications where the following characteristics are of paramount concern:

- Mechanical strength
- Excellent electrical properties

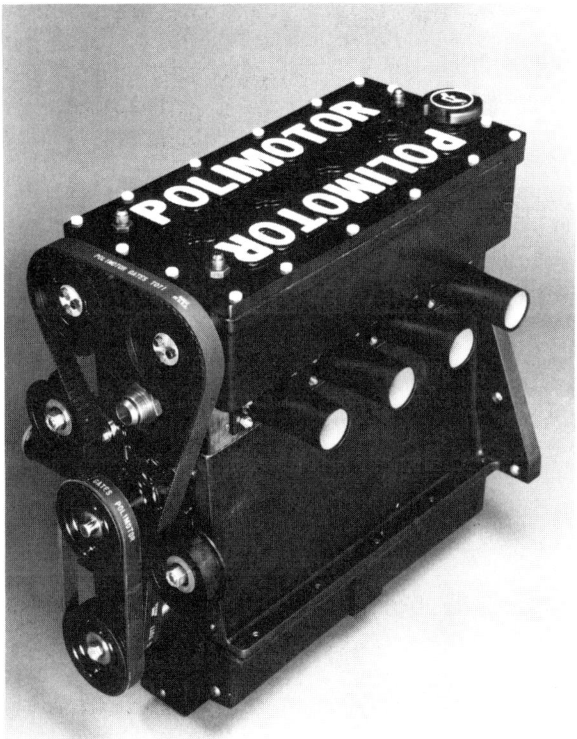

Figure 2. Thermoset molded engine block.

- High thermal resistance
- High chemical resistance
- Dimensional stability
- Economical production
- Broad product base
- Solid engineering-based industry

A relatively recent trend to produce alloys of thermosetting and thermoplastic resin systems in a polymer–matrix composite (PMC) has brought forth a new group of compounds exhibiting the best characteristics of both resin systems:

- Superior strength:weight ratios
- Greater overall toughness
- Improved moldability
- Ability to design, create, and mold complex configurations
- Excellent aesthetics

The author, in his early years, decided to become involved in "plastics" because it gave the promise of an industry whose surface had hardly been scratched. Almost half a century later, the same perception is still valid. It seems as though hardly a month goes by without a new and exciting plastic resin/reinforcement system coming on the market. The extent to which the blending or alloying of thermosetting and thermoplastic resin systems is taking place gives great hope for the future of the entire plastics industry as well as a grand challenge to the minds of all participants, whether involved in compounding, molding, mold designing, product designing, production engineering, sales, engineering, or marketing.

It has often been predicted that the thermosetting phase of the plastics industry was obsolete, but this has not been true nor will it ever be entirely true. As long as there are applications for molded part performance, under loads in excess of 264 psi, temperatures in excess of 212°F, and over extended time periods, thermosets will be the compound choice.

REFERENCE

1. Rogers Corp. MMD, Manchester, CT.

APPENDIX

CONVERSION TABLES

LINEAR MEASURE

12 inches = 1 foot
3 feet = 1 yard = 36 inches
5½ yards = 1 rod or pole = 16½ feet
40 rods = 1 furlong = 220 yards = 660 feet = ⅛ mile
8 furlongs = 1 statute mile = 1,760 yards = 5,280 feet
3 miles = 1 league = 5,280 yards = 15,840 feet

SQUARE MEASURE

144 square inches = 1 square foot
9 square feet = 1 square yard = 1,296 square inches
30¼ square yards = 1 square rod = 272¼ square feet
160 square rods = 1 acre = 4,840 square yards
640 acres = 1 square mile = 3,097,600 square yards

CUBIC MEASURE

1728 cubic inches = 1 cubic foot
27 cubic feet = 1 cubic yard
144 cubic inches = 1 board foot
128 cubic feet = 1 cord

LIQUID MEASURE

4 gills = 1 pint
2 pints = 1 quart = 8 gills
4 quarts = 1 gallon = 8 pints = 32 gills
31½ gallons = 1 barrel = 126 quarts
2 barrels = 1 hogshead = 63 gallons = 252 quarts

DRY MEASURE

2 pints = 1 quart
8 quarts = 1 peck = 16 pints
4 pecks = 1 bushel = 32 quarts = 64 pints
105 quarts = 1 barrel (for fruits, vegetables, and other dry commodities) = 7,056 cubic inches

CIRCULAR MEASURE

60 seconds (″) = 1 minute (′)
60 minutes = 1 degree (°)
90 degrees = 1 quadrant
4 quadrants = 1 circle of circumference

NAUTICAL MEASURE

6 feet = 1 fathom
100 fathoms = 1 cable's length (ordinary) = 608 Ft. (Br.) = 607.61 Ft. (U.S.)
120 fathoms = 1 cable's length (U.S. Navy)
10 cable's lengths = 1 nautical mile = 6,080 Ft. (Br.) = 6,076.1033 Ft. (U.S.)
1 nautical mile = 1.1508 statute miles
3 nautical miles = 1 league (marine)
60 nautical miles = 1 degree (of a terrestrial great circle)

AVOIRDUPOIS WEIGHT

$27\frac{11}{32}$ grains = 1 dram
16 drams = 1 ounce = $437\frac{1}{2}$ grains
16 ounces = 1 pound = 256 drams = 7,000 grains
100 pounds = 1 hundredweight = 1,600 ounces
20 hundredweight = 1 ton = 2,000 pounds
112 pounds = 1 long hundredweight
20 long hundredweight = 1 long ton = 2,240 pounds

TROY WEIGHT

24 grains = 1 pennyweight
20 pennyweights = 1 ounce = 480 grains
12 ounces = 1 pound = 240 pennyweights = 5,760 grains

APOTHECARIES' WEIGHT

20 grains = 1 scruple
3 scruples = 1 dram = 60 grains
8 drams = 1 ounce = 24 scruples = 480 grains
12 ounces = 1 pound = 96 drams = 288 scruples = 5,760 grains

Specific Gravities of Materials

Material	Specific gravity
Aluminum	2.7
Asbestos	2.0–2.8
Brass	8.5
Brick	1.4–2.2
Celluloid	1.4
Cellulose acetate	1.27–1.63
Cold molded compound	2.0
Copper	8.8
Rogers DAP/DAIP	1.36–1.80
Glass	2.4–5.9
Hard rubber	1.1–1.4
Hard wood	0.65–1.23
India rubber	0.91
Iron	7.0–7.9
Ivory	1.83–1.92
Lead	11.3
Marble	2.6–2.8
Methyl methacrylate	1.18
Rogers phenolics	1.25–2.08
Melamine formaldehyde	1.47–1.52
Polystyrene	1.05–1.07
Porcelain	2.5
Shellac	1.6
Soft wood	0.38–0.92
Steel	7.6–7.8
Water	1.0
Zinc	7.1

Conversion of Specific Gravity to Grams per Cubic Inch[a]
16.39 × Specific Gravity = Grams/In.

Specific gravity	Grams/In.	Specific gravity	Grams/In.
1.20	19.7	1.82	29.8
1.22	20.0	1.84	30.2
1.24	20.3	1.86	30.5
1.26	20.7	1.88	30.8
1.28	21.0	1.90	31.1
1.30	21.3	1.92	31.5
1.32	21.6	1.94	31.8
1.34	22.0	1.96	32.1
1.36	22.3	1.98	32.5
1.38	22.6	2.00	32.8
1.40	22.9	2.02	33.1
1.42	23.3	2.04	33.4
1.44	23.6	2.06	33.8
1.46	23.9	2.08	34.1
1.48	24.3	2.10	34.4
1.50	24.6	2.12	34.7
1.52	24.9	2.14	35.1
1.54	25.2	2.16	35.4
1.56	25.6	2.18	35.7
1.58	25.9	2.20	36.1
1.60	26.2	2.22	36.4
1.62	26.6	2.24	36.7
1.64	26.9	2.26	37.0
1.66	27.2	2.28	37.4
1.68	27.5	2.30	37.7
1.70	27.9	2.32	38.0
1.72	28.2	2.34	38.4
1.74	28.5	2.36	38.7
1.76	28.8	2.38	39.0
1.78	29.2	2.40	39.3
1.80	29.5		

[a]To determine the cost/cu./in.:
price/lb. × sp. gravity × .03163.
$1.32 × 1.76 × .03163 = $0.09/cu./in.

Diameters and Areas of Circles

Diameter (in.)	Area	Diameter (in.)	Area	Diameter (in.)	Area	Diameter (in.)	Area
1/64	.00019	7/8	2.7612	11/16	17.257	7/8	61.862
1/32	.00077	15/16	2.9483	3/4	17.721		
3/64	.00173			13/16	18.190	9–	63.617
1/16	.00307	2–	3.1416	7/8	18.665	1/8	65.397
3/32	.00690	1/16	3.3410	15/16	19.147	1/4	67.201
1/8	.01227	1/8	3.5466			3/8	69.029
5/32	.01917	3/16	3.7583	5–	19.635	1/2	70.882
3/16	.02761	1/4	3.9761	1/16	20.129	5/8	72.760
7/32	.03758	5/16	4.2000	1/8	20.629	3/4	74.662
1/4	.04909	3/8	4.4301	3/16	21.125	7/8	76.589
9/32	.06213	7/16	4.6664	1/4	21.648		
5/16	.07670	1/2	4.9087	5/16	22.166	10–	78.540
11/32	.09281	9/16	5.1572	3/8	22.691	1/8	80.516
3/8	.11045	5/8	5.4119	7/16	23.211	1/4	82.516
13/32	.12962	11/16	5.6727	1/2	23.758	3/8	84.541
7/16	.15033	3/4	5.9396	9/16	24.301	1/2	86.590
15/32	.17257	13/16	6.2126	5/8	24.850	5/8	88.664
1/2	.19635	7/8	6.4918	11/16	25.406	3/4	90.763
17/32	.22165	15/16	6.7771	3/4	25.967	7/8	92.886
9/16	.24850			13/16	26.535		
19/32	.27688	3–	7.0686	7/8	27.109	11–	95.033
5/8	.30680	1/16	7.3662	15/16	27.688	1/2	103.87
21/32	.33824	1/8	7.6699				
11/16	.37122	3/16	7.9798	6–	28.274	12–	113.10
23/32	.40574	1/4	8.2958	1/8	29.465	1/2	122.72
3/4	.44179	5/16	8.6179	1/4	30.680		
25/32	.47937	3/8	8.9462	3/8	31.919	13–	132.73
13/16	.51849	7/16	9.2806	1/2	33.183	1/2	143.14
27/32	.55914	1/2	9.6211	5/8	34.472		
7/8	.60132	9/16	9.9678	3/4	35.785	14–	153.94
29/32	.64504	5/8	10.321	7/8	37.122	1/2	165.13
15/16	.69029	11/16	10.680				
31/32	.73708	3/4	11.045	7–	38.485	15–	176.71
		13/16	11.416	1/8	39.871	1/2	188.69
1–	.7854	7/8	11.793	1/4	41.282		
1/16	.8866	15/16	12.177	3/8	42.718	16–	201.06
1/8	.9940			1/2	44.179	1/2	213.82
3/16	1.1075	4–	12.566	5/8	45.664		
1/4	1.2272	1/16	12.962	3/4	47.173	17–	226.98
5/16	1.3530	1/8	13.364	7/8	48.707	1/2	240.53
3/8	1.4849	3/16	13.772				
7/16	1.6230	1/4	14.186	8–	50.265	18–	254.47
1/2	1.7671	5/16	14.607	1/8	51.849	1/2	268.80
9/16	1.9175	3/8	15.033	1/4	53.456	19–	283.53
5/8	2.0739	7/16	15.466	3/8	55.088	1/2	298.65
11/16	2.2465	1/2	15.904	1/2	56.745		
3/4	2.4053	9/16	16.349	5/8	58.426	20–	314.16
13/16	2.5802	5/8	16.800	3/4	60.132	1/2	330.06

Steam Temperature versus Gauge Pressure

Gauge pressure (lb)	Temperature (°F)
50	297.5
55	302.4
60	307.1
65	311.5
70	315.8
75	319.8
80	323.6
85	327.4
90	331.1
95	334.3
100	337.7
105	341.0
110	344.0
115	347.0
120	350.0
125	353.0
130	356.0
135	358.0
140	361.0
145	363.0
150	365.6
155	368.0
160	370.3
165	372.7
170	374.9
175	377.2
180	379.3
185	381.4
190	383.5
195	385.7
200	387.5

Weight of 1000 Pieces in Pounds Based on Weight of One Piece in Grams[a]

Weight per piece in grams	Weight per 1000 pieces in pounds	Weight per piece in grams	Weight per 1000 pieces in pounds
1	2.2	51	112.3
2	4.4	52	114.5
3	6.6	53	116.7
4	8.8	54	118.9
5	11.0	55	121.1
6	13.2	56	123.3
7	15.4	57	125.5
8	17.6	58	127.7
9	19.8	59	129.9
10	22.0	60	132.1
11	24.2	61	134.3
12	26.4	62	136.5
13	28.6	63	138.7
14	30.8	64	140.9
15	33.0	65	143.1
16	35.2	66	145.3
17	37.4	67	147.5
18	39.6	68	149.7
19	41.8	69	151.9
20	44.0	70	154.1
21	46.2	71	156.3
22	48.4	72	158.5
23	50.6	73	160.7
24	52.8	74	162.9
25	55.0	75	165.1
26	57.2	76	167.4
27	59.4	77	169.6
28	61.6	78	171.8
29	63.8	79	174.0
30	66.0	80	176.2
31	68.2	81	178.4
32	70.4	82	180.6
33	72.6	83	182.8
34	74.8	84	185.0
35	77.0	85	187.2
36	79.2	86	189.4
37	81.4	87	191.6
38	83.7	88	193.8
39	85.9	89	196.0
40	88.1	90	198.2
41	90.3	91	200.4
42	92.5	92	202.6
43	94.7	93	204.8
44	96.9	94	207.0
45	99.1	95	209.2
46	101.3	96	211.4
47	103.5	97	213.6
48	105.7	98	215.8
49	107.9	99	218.0
50	110.1	100	220.2

[a]Equivalent weights:
1 gram = .0353 oz.
.0625 pounds = 1 ounce = 28.3 grams
454 grams = 1 pound

Length Equivalents[a]: Millimeters to Inches

Millimeters	Inches	Millimeters	Inches	Millimeters	Inches
1	.03937	34	1.33860	67	2.63779
2	.07874	35	1.37795	68	2.67716
3	.11811	36	1.41732	69	2.71653
4	.15748	37	1.45669	70	2.75590
5	.19685	38	1.49606	71	2.79527
6	.23622	39	1.53543	72	2.83464
7	.27559	40	1.57480	73	2.87401
8	.31496	41	1.61417	74	2.91338
9	.35433	42	1.65354	75	2.95275
10	.39370	43	1.69291	76	2.99212
11	.43307	44	1.73228	77	3.03149
12	.47244	45	1.77165	78	3.07086
13	.51181	46	1.81102	79	3.11023
14	.55118	47	1.85039	80	3.14960
15	.59055	48	1.88976	81	3.18897
16	.62992	49	1.92913	82	3.22834
17	.66929	50	1.96850	83	3.26771
18	.70866	51	2.00787	84	3.30708
19	.74803	52	2.04724	85	3.34645
20	.78740	53	2.08661	86	3.38582
21	.82677	54	2.12598	87	3.42519
22	.86614	55	2.16535	88	3.46456
23	.90551	56	2.20472	89	3.50393
24	.94488	57	2.24409	90	3.54330
25	.98425	58	2.28346	91	3.58267
26	1.02362	59	2.32283	92	3.62204
27	1.06299	60	2.36220	93	3.66141
28	1.10236	61	2.40157	94	3.70078
29	1.14173	62	2.44094	95	3.74015
30	1.18110	63	2.48031	96	3.77952
31	1.22047	64	2.51968	97	3.81889
32	1.25984	65	2.55905	98	3.85826
33	1.29921	66	2.59842	99	3.89763
				100	3.93700

[a]Volume equivalents:
1 c.c. = .061 cu. in.
1 cu. in. = 16.387 c.c.

Decimal Equivalents of Fractions of One Inch

1/64	.015625	17/64	.265625	33/64	.515625	49/64	.765625
1/32	.031250	9/32	.281259	17/32	.531250	25/32	.781250
3/64	.046875	19/64	.296875	35/64	.546875	51/64	.796875
1/16	.062500	5/18	.312500	9/16	.562500	13/16	.812500
5/64	.078125	21/64	.328125	37/64	.578125	53/64	.828125
3/32	.093750	11/32	.343750	19/32	.593750	27/32	.843750
7/64	.109375	23/64	.359375	39/64	.609375	55/64	.859375
1/8	.125000	3/8	.375000	5/8	.625000	7/8	.875000
9/64	.140625	25/64	.390625	41/64	.640625	57/64	.890625
5/32	.156250	13/32	.406250	21/32	.656250	29/32	.906250
11/64	.171875	27/64	.421875	43/64	.671875	59/64	.890625
3/16	.187500	7/16	.437500	11/16	.687500	15/16	.937500
13/64	.203125	29/64	.453125	45/64	.703125	61/64	.953125
7/32	.218750	15/32	.468750	23/32	.718750	31/32	.968750
15/64	.234375	31/64	.484375	47/64	.734375	63/64	.984375
1/4	250000	1/2	.500000	3/4	.750000	1	1.000000

Temperature Conversion Table[a]

°C		°F	°C		°F	°C		°F
-40.0	-40	-40.0	-6.7	20	68.0	15.6	60	140.0
-38.9	-38	-36.4	-6.1	21	69.8	16.1	61	141.8
-37.8	-36	-32.8	-5.6	22	71.6	16.7	62	143.6
-36.7	-34	-29.2	-5.0	23	73.4	17.2	63	145.4
-35.6	-32	-25.6	-4.4	24	75.2	17.8	64	147.2
-34.4	-30	-22.0	-3.9	25	77.0	18.3	65	149.0
-33.3	-28	-18.4	-3.3	26	78.8	18.9	66	150.8
-32.2	-26	-14.8	-2.8	27	80.6	19.4	67	152.6
-31.1	-24	-11.2	-2.2	28	82.4	20.0	68	154.4
-30.0	-22	- 7.6	-1.7	29	84.2	20.6	69	156.2
-28.9	-20	- 4.0	-1.1	30	86.0	21.1	70	158.0
-27.8	-18	- 0.4	-0.6	31	87.8	21.7	71	159.8
-26.7	-16	3.2	0.0	32	89.6	22.2	72	161.6
-25.6	-14	6.8	0.6	33	91.4	22.8	73	163.4
-24.4	-12	10.4	1.1	34	93.2	23.3	74	165.2
-23.3	-10	14.0	1.7	35	95.0	23.9	75	167.0
-22.2	- 8	17.6	2.2	36	96.8	24.4	76	168.8
-21.1	- 6	21.2	2.8	37	98.6	25.0	77	170.6
-20.0	- 4	24.8	3.3	38	100.4	25.6	78	172.4
-18.9	- 2	28.4	3.9	39	102.2	26.1	79	174.2
-17.8	0	32.0	4.4	40	104.0	26.7	80	176.0
-17.2	1	33.8	5.0	41	105.8	27.2	81	177.8
-16.7	2	35.6	5.6	42	107.6	27.8	82	179.6
-16.1	3	37.4	6.1	43	109.4	28.3	83	181.4
-15.6	4	39.2	6.7	44	111.2	28.9	84	183.2
-15.0	5	41.0	7.2	45	113.0	29.4	85	185.0
-14.4	6	42.8	7.8	46	114.8	30.0	86	186.8
-13.9	7	44.6	8.3	47	116.6	30.6	87	188.6
-13.3	8	46.4	8.9	48	118.4	31.1	88	190.4
-12.8	9	48.2	9.4	49	120.2	31.7	89	192.2
-12.2	10	50.0	10.0	50	122.0	32.2	90	194.0
-11.7	11	51.8	10.6	51	123.8	32.8	91	195.8
-11.1	12	53.6	11.1	52	125.6	33.3	92	197.6
-10.6	13	55.4	11.7	53	127.4	33.9	93	199.4
-10.0	14	57.2	12.2	54	129.2	34.4	94	201.2
- 9.4	15	59.0	12.8	55	131.0	35.0	95	203.0
- 8.9	16	60.8	13.3	56	132.8	35.6	96	204.8
- 8.3	17	62.6	13.9	57	134.6	36.1	97	206.6
- 7.8	18	64.4	14.4	58	136.4	36.7	98	208.4
- 7.2	19	66.2	15.0	59	138.2	37.2	99	210.2

[a]Numbers in the center column refer to temperature, either centigrade or fahrenheit, which is to be converted into the other scale. To convert from fahrenheit to centigrade find equivalent temperature in left-hand column marked C; in like manner find equivalent temperature in right-hand column when converting from centigrade to fahrenheit. Example: 60°F is 15.6°C; 60°C is 140°F.

$$°C = \frac{(5 \times °F) - 160}{9} \qquad °F = \frac{(9 \times °C) + 160}{5}$$

Temperature Conversion Table (*cont.*)

°C		°F	°C		°F	°C		°F
37.8	100	212.0	82.2	180	356.0	137.8	280	536.0
38.9	102	215.6	83.3	182	359.6	140.6	285	545.0
40.0	104	219.2	84.4	184	363.2	143.3	290	554.0
41.1	106	222.8	85.6	186	366.8	146.1	295	563.0
42.2	108	226.4	86.7	188	370.4	148.9	300	572.0
43.3	110	230.0	87.8	190	374.0	151.7	305	581.0
44.4	112	233.6	88.9	192	377.6	154.4	310	590.0
45.6	114	237.2	90.0	194	381.2	157.2	315	599.0
46.7	116	240.8	91.1	196	384.8	160.0	320	608.0
47.8	118	244.4	92.2	198	388.4	162.8	325	617.0
48.9	120	248.0	93.3	200	392.0	165.6	330	626.0
50.0	122	251.6	94.4	202	395.6	168.3	335	635.0
51.1	124	255.2	95.6	204	399.2	171.1	340	644.0
52.2	126	258.8	96.7	206	402.8	173.9	345	653.0
53.3	128	262.4	97.8	208	406.4	176.7	350	662.0
54.4	130	266.0	98.9	210	410.0	179.4	355	671.0
55.6	132	269.6	100.0	212	413.6	182.2	360	680.0
56.7	134	273.2	101.1	214	417.2	185.0	365	689.0
57.8	136	276.8	102.2	216	420.8	188.2	370	698.0
58.9	138	280.4	103.3	218	424.4	190.6	375	707.0
60.0	140	284.0	104.4	220	428.0	193.3	380	716.0
61.1	142	287.6	105.6	222	431.6	196.1	385	725.0
62.2	144	291.2	106.7	224	435.2	198.9	390	734.0
63.3	146	294.8	107.8	226	438.8	201.7	395	743.0
64.4	148	298.4	108.9	228	442.4	204.4	400	752.0
65.6	150	302.0	110.0	230	446.0	210.0	410	770.0
66.7	152	305.6	111.1	232	449.6	215.6	420	788.0
67.8	154	309.2	112.2	234	453.2	221.1	430	806.0
68.9	156	312.8	113.3	236	456.8	226.7	440	824.0
70.0	158	316.4	114.4	238	460.4	232.2	450	842.0
71.1	160	320.0	115.6	240	464.0	237.8	460	860.0
72.2	162	323.6	116.7	242	467.6	243.3	470	878.0
73.3	164	327.2	117.8	244	471.2	248.9	480	896.0
74.4	166	330.8	118.3	245	473.0	254.5	490	914.0
75.6	168	334.4	121.1	250	482.0	260.0	500	932.0
76.7	170	338.0	123.9	255	491.0	265.6	510	950.0
77.8	172	341.6	126.7	260	500.0	271.1	520	968.0
78.9	174	345.2	129.4	265	509.0	276.7	530	986.0
80.0	176	348.8	132.2	270	518.0	282.3	540	1004.0
81.1	178	352.4	135.0	275	527.0	287.8	550	1022.0

Standard Draft Angles

Depth	¼°	½°	1°	1½°	2°	2½°	3°	5°	7°	8°	10°	12°	15°	Depth
1/32	.0001	.0003	.0005	.0008	.0011	.0014	.0016	.0027	.0038	.0044	.0055	.0066	.0084	1/32
1/16	.0003	.0006	.0011	.0016	.0022	.0027	.0033	.0055	.0077	.0088	.0110	.0133	.0168	1/16
3/32	.0004	.0008	.0016	.0025	.0033	.0041	.0049	.0082	.0115	.0132	.0165	.0199	.0251	3/32
1/8	.0005	.0010	.0022	.0033	.0044	.0055	.0066	.0109	.0153	.0176	.0220	.0266	.0335	1/8
3/16	.0008	.0016	.0033	.0049	.0065	.0082	.0098	.0164	.0230	.0263	.0331	.0399	.0502	3/16
1/4	.0011	.0022	.0044	.0066	.0087	.0109	.0131	.0219	.0307	.0351	.0441	.0531	.0670	1/4
5/16	.0014	.0027	.0055	.0082	.0109	.0137	.0164	.0273	.0384	.0439	.0551	.0664	.0837	5/16
3/8	.0016	.0033	.0065	.0098	.0131	.0164	.0197	.0328	.0460	.0527	.0661	.0797	.1005	3/8
7/16	.0019	.0038	.0076	.0115	.0153	.0191	.0229	.0383	.0537	.0615	.0771	.0930	.1172	7/16
1/2	.0022	.0044	.0087	.0131	.0175	.0218	.0262	.0438	.0614	.0703	.0882	.1063	.1340	1/2
5/8	.0027	.0054	.0109	.0164	.0218	.0273	.0328	.0547	.0767	.0878	.1102	.1329	.1675	5/8
3/4	.0033	.0065	.0131	.0196	.0262	.0328	.0393	.0656	.0921	.1054	.1322	.1595	.2010	3/4
7/8	.0038	.0076	.0153	.0229	.0306	.0382	.0459	.0766	.1074	.1230	.1543	.1860	.2345	7/8
1	.0044	.0087	.0175	.0262	.0349	.0437	.0524	.0875	.1228	.1405	.1763	.2126	.2680	1
1¼	.0055	.0109	.0218	.0327	.0437	.0546	.0655	.1094	.1535	.1756	.2204	.2657	.3349	1¼
1½	.0064	.0131	.0262	.0393	.0524	.0655	.0786	.1312	.1842	.2108	.2645	.3188	.4019	1½
1¾	.0076	.0153	.0305	.0458	.0611	.0764	.0917	.1531	.2149	.2460	.3085	.3720	.4689	1¾
2	.0087	.0175	.0349	.0524	.0698	.0873	.1048	.1750	.2456	.2810	.3527	.4251	.5359	2
Depth	¼°	½°	1°	1½°	2°	2½°	3°	5°	7°	8°	10°	12°	15°	Depth

INDEX